THE RAT
AS A SMALL MAMMAL

by

H. G. Q. ROWETT, M.A., F.R.S.A., M.I.Biol.

SENIOR LECTURER AT PLYMOUTH POLYTECHNIC

THIRD EDITION

JOHN MURRAY
FIFTY ALBEMARLE STREET LONDON

INTRODUCTION

MAMMALS are **tetrapod vertebrates**, i.e. four-limbed animals with backbones. The diagnostic feature of the group as a whole is the possession of **mammary glands** which produce milk on which the young are fed during the early stages of their lives. With few exceptions, mammals have **hairy skin** and are **viviparous**, retaining the young in the body of the mother during the early stages of development instead of laying eggs. They are **warm-blooded**, having an approximately constant body temperature irrespective of external conditions and they breathe by means of **lungs**, which, with the heart, lie in a compartment known as the **thorax**, separated from the other internal organs by a muscular **diaphragm**.

There are numerous subdivisions of the *Mammalia* known as orders. Each of these has its own characteristics. In older classifications rats, guinea pigs and rabbits are all included in the order *Rodentia*, i.e. animals with rodent or gnawing front or incisor teeth. This order was then divided into two sub-orders, the *Simplicidentata* (rats, guinea pigs) with a single pair of upper incisors and the *Duplicidentata* (rabbits, hares) with two pairs of upper incisors. There are, however, a number of other differences affecting the teeth, skull and soft parts so that the *Duplicidentata* have now been removed to a new order, the *Lagomorpha*, and the general similarity of the incisor teeth is attributed to parallel adaptation to the gnawing habit.

Rats, guinea pigs and rabbits are all small enough to be convenient for dissection, though large enough to show the parts without difficulty. All are easy to keep and breed and, until recently, the rabbit was most usually studied as a general mammalian type. Increasing price and regulations regarding the transport of the living animals have encouraged many to change to the study of the rat, a change with the additional advantages of the vast amount of knowledge of rat physiology and the greater similarity in many ways to man himself. Throughout this book care has been taken to emphasise those features which are peculiar to rats and do not occur in mammals in general or in man. It is hoped that this treatment will provide a useful introduction to medical studies as well as being suitable for biology and zoology courses. There is more detail than is needed by many students at school level, but this is given in order to produce a complete picture and draw together data from many sources for the benefit of those who require it. Notes are included on the habits of rats and on the methods of keeping and breeding them under ordinary school conditions. Wherever scales or diagrams are given they are multiples of a standard represented by the average size of an adult albino rat.

Acknowledgements

I wish to express my thanks to all those who have helped with this work, including Mr. G. Porter and Miss M. P. Williams who set me on the right way; Dr. S. Smith for advice on reproduction and embryology; Major C. W. Hume, M.C., for helpful criticism of Appendix I; Mr. N. Bateman for the data for Appendix II; and most especially Miss L. W. Turpitt for her advice throughout.

The warm reception of the first edition of this book has led to an early reprint and the opportunity has been taken to make a number of changes in the text in order to bring it as up-to-date as possible. I wish to thank all those who have written to me regarding points from their special research, especially Dr. E. W. Bentley, Mr. C. R. Kennedy, Mr. C. G. Gardiner, Dr. M. E. Varley, and Dr. J. F. Frazer, and to thank Dr. J. M. Barnes for permission to use his silver impregnation photograph for the preparation of the new Fig. 55. I further add my thanks to Mr. B. Dunne for his help with the revision leading to the yet more up-to-date third edition.

H.G.Q.R.

CONTENTS

© H. G. Q. Rowett 1960

First edition 1957
Second edition 1960
Reprinted 1962, 1965 (revised), 1968
Third edition 1974

0 7195 2996 4

Printed in Great Britain by Jarrold & Sons Ltd, Norwich, and published by John Murray (Publishers) Ltd.

BRITISH RATS

THE rat belongs to the largest order of mammals, the *Rodentia*, which also includes the mice, guinea pigs, squirrels, marmots, beavers, and jerboas and many other less familiar types. Rats and mice form the sub-family *Muridae* of the family *Muroidea*, which also includes the voles and lemmings.

There are two species of rat found wild in Britain, the black rat, *Rattus rattus* and the brown rat, *Rattus norvegicus*. In practice they cannot be distinguished by colour because in both cases this is extremely variable. The black rat may have the brown back and light belly characteristic of the brown rat, while the brown rat is often dark grey. The black rat is, however, smaller and less heavily built than the brown rat and there are a number of differences in behaviour.

Neither of these species of rat is indigenous in Britain. The black rat came originally from southern Asia and spread through Europe during the Middle Ages, reaching England at latest in the thirteenth century and possibly much earlier. The brown rat came originally from central Asia and is believed to have been brought to England in 1730 on the occasion of a visit from the Russian fleet. Both species of rat occur all over the world, having been trans-shipped accidentally with various cargoes and having established themselves wherever there are human settlements. They do not come into competition with the various native rats because the latter will not enter either houses or ships, but when they are in competition with one another the very ferocious brown rat tends to exterminate the milder black rat.

The black rat is the commoner species in the tropics. In colder climates it is limited to human habitations. It is therefore typically a town-dweller and lives chiefly in old cellars and warehouses. It is also common on ships. In its native habitat it is somewhat **arboreal** and therefore, when frightened, it tends to climb upwards to hide. It can run along telegraph wires and over roofs and often enters buildings through the sky-lights. It is an unwilling swimmer and is rarely found in sewers.

The brown rat can stand more cold than the black rat and is therefore commoner in temperate regions, where it does not need to seek human shelter, though it often does so. It makes burrows and is generally less agile and more **terrestrial** than the black rat. It usually enters buildings through cracks in the flooring and through drains. It is a good swimmer.

Both types of rat are subject to **bubonic plague**, but the black rat is much more susceptible than the brown rat, so that it is much the more frequent carrier of the disease. Plague is transmitted by fleas, particularly the rat flea, *Xenopsylla cheopsis*. This flea does not normally attack human beings, but when rats are dying of plague it may seek man as an alternative host and transmit plague to him. Epidemic amongst rats thus precedes epidemic amongst men, and the rat population acts as a reservoir for the disease. The "Black Death" of the fourteenth century was spread by rats, and there were many subsequent outbreaks of plague in Europe, including the great plague of London which was probably spread by the rats infesting the old city before the great fire in 1666. After the end of the seventeenth century, plague gradually disappeared from Europe, though still occurring in Asia and Africa. As a precaution against the infection of the native British rats through rats from abroad, special conical rat guards are placed on the mooring ropes of all ocean-going ships and unwatched gangways are never left down at night. An anti-plague vaccine is available for those who must travel to plague-ridden places.

Rats also carry two of the many forms of **typhus**. **Scrub typhus** is transmitted by the blood-sucking larval stages of certain mites. The virus of the disease can remain in the mites during their non-parasitic nymphal and adult stages and be passed on to the eggs; thus the rats are not essential to the existence of the virus but act as reservoirs from which more mites may become infected. **Endemic** or **murine typhus** is transmitted from rat to rat by fleas, lice and mites and from rat to man by fleas or contamination of food. Rat lice do not bite man.

Food poisoning or **gastro-enteritis**, caused by bacteria of the Salmonella group, may be transmitted directly by rats through contamination of food.

Weil's disease, a form of infective jaundice, is caused by a spirochaete found in the urine of rats. It is commonest among sewer workers, to whom it is usually transmitted through contact of a skin abrasion with rat urine, urine contaminated surfaces or bodies of dead rats. **Rat bite fever** may result from a bite or scratch from an infected rat.

In addition to their danger as possible carriers of disease, rats do a great deal of damage by their **omnivorous** habits. They destroy crops, poultry and stored products, frequently hoarding food and even carrying away eggs for this purpose. The brown rat is more of a scavenger and less selective than the black rat but both are extremely conservative in their choice of food. When attempting to destroy rats, warfarin and other chronic poisons should be used so that the rat does not associate the poison with the food. The more acute types of poison may cause the rat to become bait shy before it has taken enough to be seriously harmed. If traps are used they should be left baited but unset for several days till the rats are accustomed to them.

The **fecundity** of rats is so great that if any are left they can rapidly repopulate the region. A healthy female rat will produce about 30 offspring a year by having 4–5 families of 4–10 young. The young start to breed when they are between 3 and 6 months old. Hence rats are in general most suitable for genetical and other experiments. The physiology of rats is remarkably similar to that of the human being, so that they have been much used in feeding tests, in estimating the potency of vitamin preparations and in work with new drugs. The domesticated rat which is most usually used is an albino variety. It is derived from and will interbreed with the brown rat, *Rattus norvegicus*, but not with the black rat. It is smaller in size than its wild relative and mild natured so that it can be handled with complete safety. Some notes on its husbandry are given in Appendix I.

×Pg26

EXTERNAL APPEARANCE

SHAPE

THE rat has a long cylindrical body, long, thin tail and very short legs. This shape is well suited to running in narrow burrows or squeezing through small holes.

The **head** has a pointed snout with two slit-like **nostrils**, a narrow **mouth** with short lower jaw and split upper lip, two small beady **eyes** set so that they look diagonally forwards and sideways, two rounded **pinnae** and numerous long whiskers or **vibrissae.**

The **neck** is short.

The **trunk** is very little wider than the head. When in the running position, there is no conspicuous arch of the back or difference between the girth of the **thorax** and that of the **abdomen**. There is considerable flexibility of the spine so that the body can be bent well over to either side or backwards when climbing.

The **anus** is at the base of the tail. In the **female**, the **urinary** and **genital apertures** are anterior to the anus on the ventral surface of the hind end of the abdomen and there are 10–12 **nipples** forming two unevenly spaced rows, one on either side of the belly between the axillae and the groins. In the **male** the common **urinogenital aperture** is at the tip of a short retractile **penis**, behind which are two very large **scrotal sacs**. The latter extend under the base of the tail and obscure the anus from ventral view.

The **tail** is almost as long as or longer than the trunk, but very much narrower. It tapers gradually towards the hind end.

The fore and hind **limbs** are set well apart. The former are shorter than the latter, but both are well flexed so that the body is only slightly raised off the ground. Each fore-foot has four clawed **digits** and a small nodule which represents the **pollex** (thumb). Each hind-foot has five clawed digits, but the first or **hallux** is much shorter than the others. There are scaly rings on the under-surface of each digit and five scaly pads on the sole of each foot.

SIZE

The full-grown black rat and the domesticated albino variety are both about 170 mm long from the snout to the base of the tail, while the tail is over 190 mm long. The average weight of a full-grown animal is 170–200 g.

The full-grown brown rat is 190–220 mm long with a 190 mm tail. It is more heavily built than the black rat and has a proportionally shorter snout and smaller pinnae. Its average weight is about 450 g.

Fig. 1 **Rats**

× $\frac{1}{3}$

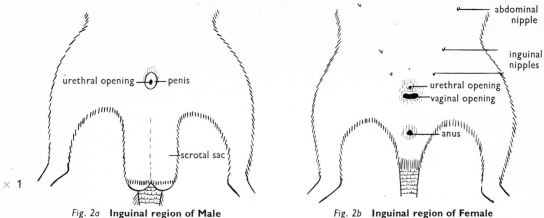

Fig. 2a **Inguinal region of Male**   Fig. 2b **Inguinal region of Female**

COLOUR

The head, neck, trunk and limbs have a dense covering of short **hairs** so that their colour is dependent on the pigmentation of the hairs. The tail has few hairs between rings of small epidermal scales. The rat has over 210 such rings while the mouse has less than 180. The pinnae and feet are naked.

The variability of the coat colour of the wild rats has already been mentioned. Domesticated rats have been bred with a wide range of coat colours and patterns. The coloration has been shown to be governed by a restricted number of genes which behave in a Mendelian manner (see Appendix II). The tail and parts of the pinnae are sometimes pigmented but the feet are usually flesh-coloured. In the albino rat, pigment is totally lacking and the fur is pure white, while the tail, pinnae, feet and eyes are pink.

SKIN

THE skin is hairy as described above. It is formed of **epidermis** and **dermis**. The epidermis is cornified, so that it is tough, while the dermis is fibrous and contains numerous blood-vessels and nerves, and small masses of fat. The skin is elastic and thus allows freedom of movement.

The **hairs** hold a layer of stagnant air near the skin and so give insulation which assists maintenance of a constant body temperature, as the animal is warm-blooded—see page 50. They can be made to "stand on end" by means of small **erector muscles.** This reaction occurs in response to cold, fear or anger, increases the insulating effect and makes the animal look larger and more ferocious.

Mammalian skin has two types of glands, the **sebaceous glands** and the **sweat glands.** Sebaceous glands produce an oily secretion which prevents the epidermis and the hairs from becoming brittle. It also makes water run off the hairs easily so that the fur does not become wet. The sweat glands produce a watery secretion which contains small amounts of salts and nitrogenous waste materials. The water evaporates and in doing so takes heat from the body, thus having a cooling effect. The rat does not have any sweat glands because the hairs are set too close together. In so small an animal such a cooling effect is in any case never required—see page 50.

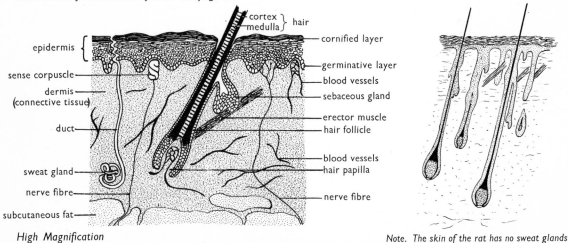

High Magnification   Note. *The skin of the rat has no sweat glands*

Fig. 3 **V.S. Mammalian skin (highly diagrammatic) with inset of actual rat skin**

SKELETON

THE plan of the skeleton is very similar in all mammals, but the details vary considerably.

As in all vertebrates the skeleton serves the following functions:

(1) It gives **support** to the soft parts and thus affects the shape of the individual.
(2) It gives **attachment** to many of the skeletal muscles which produce movements and locomotion.
(3) It gives **protection** to delicate parts such as the brain and spinal cord and the heart and lungs.

In the embryo the skeleton consists of **cartilage** but, before birth, this is largely replaced by **bone**. Most cartilages ossify in three pieces, a **diaphysis** and two **epiphyses**, and become **cartilage bones**. Bones formed directly in membranes are known as **membrane bones**. **Ossification** continues for some time after birth and some bones become fused together very late in life.

The skeleton may be considered in two parts:

1. The **axial** skeleton.
2. The **appendicular** skeleton.

1. AXIAL SKELETON

The axial skeleton includes the **skull, vertebral column, ribs** and **sternum**.

(a) THE SKULL

The skull of the rat is composed of forty-one bones.

The "cartilage" bones form the sides and floor of the cranium, the posterior part of the palate, the hyoid and the ear ossicles. They are as follows:

1 occipital (supraoccipital, basioccipital and exoccipitals fused together)
1 basisphenoid (basisphenoid and alisphenoids fused together)
1 presphenoid (presphenoid and orbitosphenoids fused together)
1 ethmoid forming the cribriform plate and part of the nasal septum
2 periotics surrounding the inner ear (mastoid and petrosal fused together)
2 palatines—part of the primitive mandibular arch
1 hyoid formed from the ventral part of the hyoid arch (basihyoid, ceratohyoids and thyrohyoids fused together; the basihyoid—body; ceratohyoids—anterior horns; thyrohyoids—posterior horns)
6 auditory ossicles—3 in each tympanic cavity:

—a malleus derived from the articular bone of the reptile jaw, and thus from the primitive mandibular arch
—an incus from the quadrate bone of the primitive mandibular arch
—a stapes from the hyomandibular bone of the hyoid arch.

The "membrane" bones form the roof of the cranium and support the face. They are as follows:

1 interparietal
2 parietals
2 frontals
2 nasals
2 lacrimals
4 turbinates
1 vomer
2 premaxillae
2 maxillae
2 jugals
2 squamosals
2 mandibles
2 tympanic bullae.

The dental formula of the rat is—I $\frac{1}{1}$; C $\frac{0}{0}$; PM $\frac{0}{0}$; M $\frac{3}{3}$ = 16.

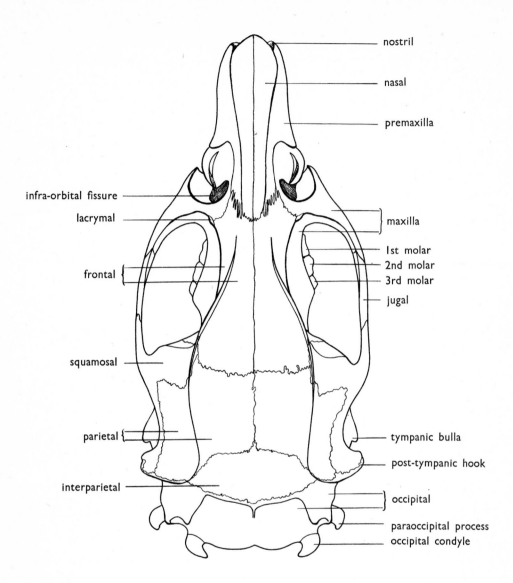

nostril

nasal

premaxilla

infra-orbital fissure

lacrymal

maxilla

1st molar
2nd molar
3rd molar
jugal

frontal

squamosal

parietal

tympanic bulla

post-tympanic hook

interparietal

occipital

paraoccipital process
occipital condyle

× 3

Fig. 4 **Skull (dorsal aspect)**

The Foramina of the Skull

The skull has a number of foramina, the most important of which are listed below with notes on their position and the structures which pass through them.

GROUP I—FORAMINA WHICH PERFORATE THE WALL OF THE CRANIUM

Foramen	Position	Structures passing through it
Foramen magnum	in occipital bone	spinal cord
Olfactory foramina	many in cribriform plate	olfactory nerves
Optic foramen	in orbitosphenoid region of presphenoid	optic nerve
Anterior lacerated foramen	between the bodies of the basisphenoid and presphenoid and the palatine	oculomotor, trochlear and abducens nerves and ophthalmic and maxillary branches of trigeminal nerve; palatine branch of internal carotid artery out of cranium
Foramen ovale	in alisphenoid region of basisphenoid	mandibular branch of trigeminal nerve
Facial canal	in periotic	facial nerve
Internal auditory meatus	in periotic	auditory nerve
Posterior lacerated foramen	between periotic and occipital	glossopharyngeal, vagus and spinal accessory nerves; internal jugular vein
Hypoglossal canal	in occipital	hypoglossal nerve
Post-glenoid foramen	between squamosal and periotic	vein from transverse sinus
Carotid canal	between basioccipital and periotic with a deep groove in the tympanic	main branch of internal carotid artery
Middle lacerated foramen	between alisphenoid region of basisphenoid and tympanic	pterygo-palatine branch of internal carotid artery out of cranium
Alisphenoid canal	in alisphenoid region of basisphenoid	palatine branch of internal carotid artery
Interpterygoid foramen	between palatine and basisphenoid	

GROUP II—FORAMINA WHICH DO NOT PERFORATE THE WALL OF THE CRANIUM

Foramen	Position	Structures passing through it
Anterior palatine foramen	between premaxilla and maxilla in the palate	naso-palatine branch of trigeminal nerve
Posterior palatine foramen	in palatine	palatine branch of trigeminal nerve
Infra-orbital fissure	in maxilla in front of orbit	maxillary branch of trigeminal nerve and the anterior deep branch of the masseter muscle
Stylomastoid foramen	between periotic and tympanic	facial nerve from bulla
Posterior petro-tympanic foramen	between periotic and tympanic close to posterior lacerated foramen	pterygo-palatine branch of internal carotid artery into bulla
Petro-tympanic fissure	between periotic and tympanic	pterygo-palatine branch of internal carotid artery through bulla into cranium
Basisphenoid canal	in basisphenoid connecting with corresponding foramen of other side within the bone	pterygoid branch of internal carotid artery
Pterygo-palatine foramen	in basisphenoid from basisphenoid canal to wall of anterior lacerated foramen	pterygoid branch of internal carotid artery
Lacrimal groove	in lacrimal	naso-lacrimal duct
Eustachian canal	in tympanic	Eustachian tube
External auditory meatus	in tympanic	outer ear passage
Mandibular foramen	in inner side of lower jaw	mandibular branch of trigeminal nerve and mandibular blood-vessels
Mental foramen	in outer side of lower jaw	mental nerve

7

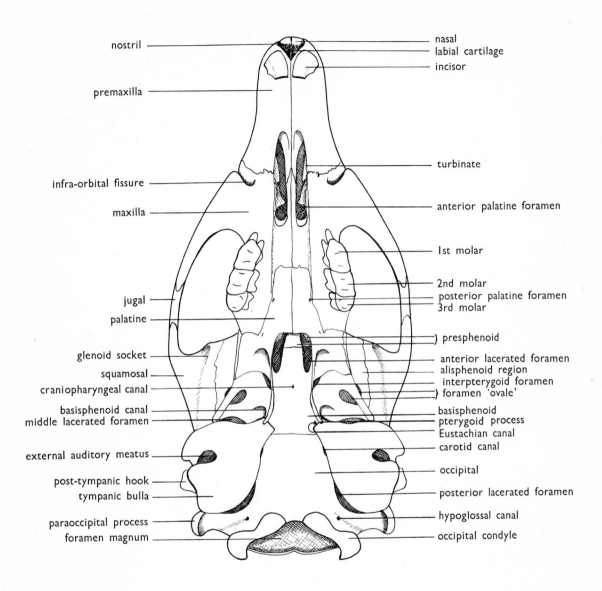

nostril — nasal / labial cartilage / incisor

premaxilla

turbinate

infra-orbital fissure

anterior palatine foramen

maxilla

1st molar

2nd molar / posterior palatine foramen / 3rd molar

jugal

palatine

presphenoid

glenoid socket

anterior lacerated foramen / alisphenoid region / interpterygoid foramen / foramen 'ovale'

squamosal

craniopharyngeal canal

basisphenoid canal

basisphenoid / pterygoid process / Eustachian canal / carotid canal

middle lacerated foramen

external auditory meatus

occipital

post-tympanic hook

posterior lacerated foramen

tympanic bulla

hypoglossal canal

paraoccipital process

occipital condyle

foramen magnum

× 3

Fig. 5 **Skull (ventral aspect)**

8

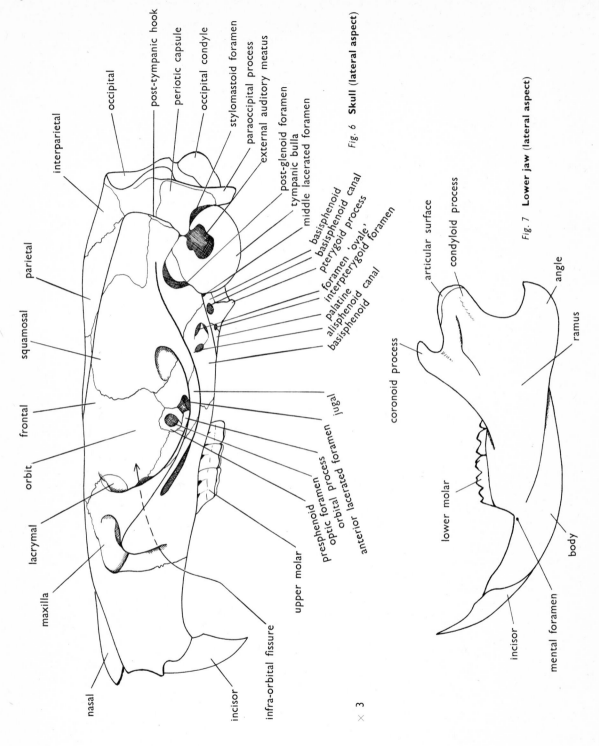

interparietal

occipital

post-tympanic hook

periotic capsule

occipital condyle

stylomastoid foramen

paraoccipital process

external auditory meatus

Fig. 6 **Skull (lateral aspect)**

post-glenoid foramen

tympanic bulla

middle lacerated foramen

parietal

basisphenoid

basisphenoid canal

pterygoid process

foramen 'ovale'

interpterygoid foramen

palatine

alisphenoid canal

basisphenoid

squamosal

frontal

orbit

jugal

lacrymal

presphenoid

optic foramen

orbital process

anterior lacerated foramen

maxilla

upper molar

nasal

incisor

infra-orbital fissure

× 3

coronoid process

articular surface

condyloid process

Fig. 7 **Lower jaw (lateral aspect)**

angle

ramus

lower molar

body

incisor

mental foramen

9

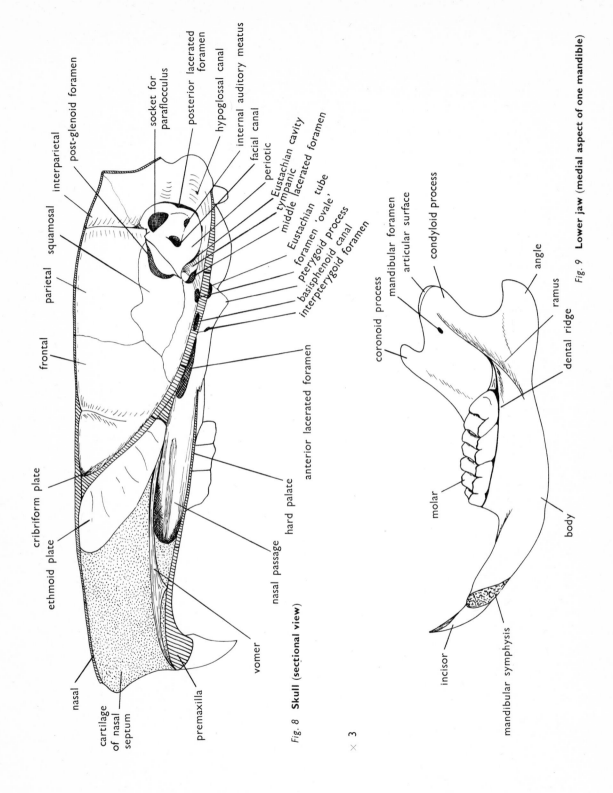

cartilage of nasal septum

nasal

ethmoid plate

cribriform plate

frontal

parietal

squamosal

interparietal

post-glenoid foramen

socket for paraflocculus

posterior lacerated foramen

hypoglossal canal

internal auditory meatus

facial canal

periotic

Eustachian cavity

tympanic

middle lacerated foramen

Eustachian tube

foramen 'ovale'

pterygoid process

basisphenoid canal

interpterygoid foramen

anterior lacerated foramen

premaxilla

vomer

nasal passage

hard palate

Fig. 8 **Skull (sectional view)**

coronoid process

mandibular foramen

articular surface

condyloid process

angle

ramus

dental ridge

body

molar

incisor

mandibular symphysis

Fig. 9 **Lower jaw (medial aspect of one mandible)**

9

3

10

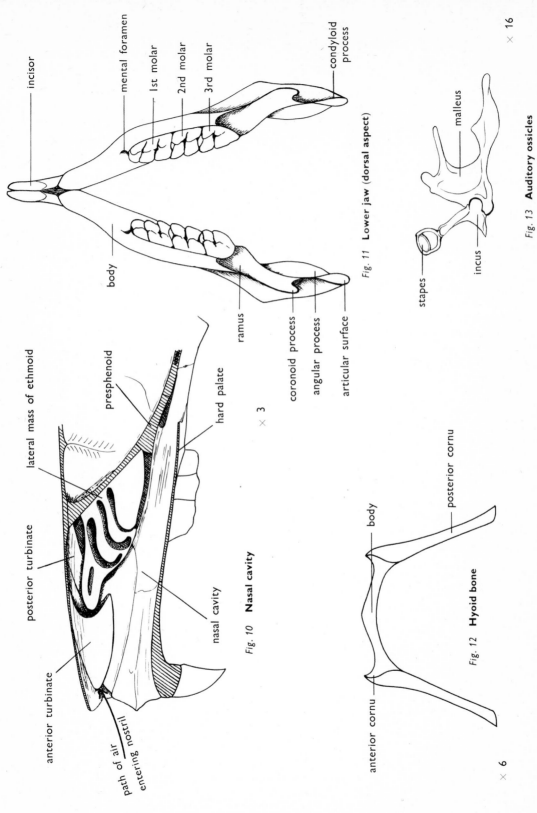

incisor

mental foramen

1st molar

2nd molar

3rd molar

condyloid process

body

ramus

coronoid process

angular process

articular surface

Fig. 11 **Lower jaw (dorsal aspect)**

× 3

malleus

stapes

incus

Fig. 13 **Auditory ossicles**

× 16

lateral mass of ethmoid

presphenoid

hard palate

posterior turbinate

nasal cavity

anterior turbinate

path of air entering nostril

Fig. 10 **Nasal cavity**

body

posterior cornu

anterior cornu

Fig. 12 **Hyoid bone**

× 6

(b) VERTEBRAL COLUMN

The vertebral column is composed of **vertebrae**. Each vertebra typically has a body or **centrum** and a **neural** arch which form the **vertebral canal** around the spinal cord, and **seven processes**. These processes are: (i) a **neural spine**; (ii) two **transverse processes**; (iii) two **anterior** and two **posterior zygapophyses**. The zygapophyses of each vertebra articulate with those of the next and allow bending but not twisting movements of the vertebral column. Between the centra there are **intervertebral discs** of cartilage, and between each neural arch and the next there are **intervertebral canals** for the passage of the spinal nerves.

The rat has fifty-seven to sixty vertebrae which may be divided into five groups: **cervical, thoracic, lumbar, sacral** and **caudal**.

The **cervical vertebrae** are characterized by the possession of **vertebrarterial canals**. These canals may be small or absent in the seventh cervical vertebra. They are formed by the fusion of small **cervical ribs** to the centra and transverse processes. Of the seven cervical vertebrae the first two are modified in the usual way as the atlas and axis. The **atlas** has articular surfaces for the occipital condyles of the skull and for the axis. It has no zygapophyses and no centrum. The **axis** has the **odontoid peg**, formed from the piece of bone which represents the missing centrum of the atlas, often not fully fused on to its own centrum and thus clearly showing its origin. The axis has no prezygapophyses but postzygapophyses are present and the neural spine is long. The other five cervical vertebrae are described as typical, but the sixth vertebra differs from the rest in the possession of enlarged cervical ribs.

The **thoracic vertebrae** are characterized by their **long neural spines** and the **articular facets** for the heads and tubercles of the free ribs. There are **thirteen** thoracic vertebrae corresponding with the number of pairs of ribs. Of these vertebrae the second has a very long neural spine which articulates with a small triangular piece of bone, which is either the epiphysis of the neural spine or a sesamoid bone.

The **lumbar vertebrae** are characterized by the possession of **anapophyses** in addition to well-developed zygapophyses, transverse processes and neural spines. The transverse processes become progressively larger in the more posterior lumbar vertebrae. There are **six** of these vertebrae and with the exception of the sacral vertebrae they are the largest in the vertebral column.

The **sacrum** is formed of **two sacral** and **two caudal vertebrae** fused together. The transverse processes of the sacral vertebrae articulate with the ilia of the pelvic girdle. The vertebrae forming the sacrum are easily identifiable by their neural spines, the fused articular processes and the intervertebral canals between them. Complete fusion occurs only in old animals. In young rats the third and fourth vertebrae of the sacrum are free and resemble caudal vertebrae except for the absence of chevron bones.

The free **caudal vertebrae** vary in number from twenty-seven to thirty. They vary in form progressively. The proximal ones show all the features of a typical vertebra and, but for the absence of anapophyses, closely resemble lumbar vertebrae. The gradual loss of all parts except the centrum is shown in the following table:

Complete vertebrae .	1 and 2
Complete except for neural spines	3 and 4
With functional prezygapophyses but non-functional postzygapophyses and without neural spine	5
With non-functional pre- and postzygapophyses, without neural spines, but with distinct transverse processes .	6–8
With zygapophyses and transverse processes reduced to small knobs and without neural spines	8–20
With centra only .	21–27 or 30

Minute **chevron bones** are present in the tail, associated with the caudal vertebrae.

The **intervertebral discs** are very well developed. In the neck region they have two vertical ridges on each surface which fit vertical grooves on the centra of the vertebrae and in the tail region they have six radially arranged depressions into which fit knobs on the centra. These ridges, knobs and grooves prevent lateral slip of the bodies of the vertebrae. The centra of the thoracic vertebrae are almost flat, but here lateral movement is prevented by the heads of the ribs. In the lumbar region the intervertebral discs are firmly fixed to the anterior surfaces of the centra and are strongly convex where they fit into the concave of the posterior surfaces of the adjacent vertebrae.

12

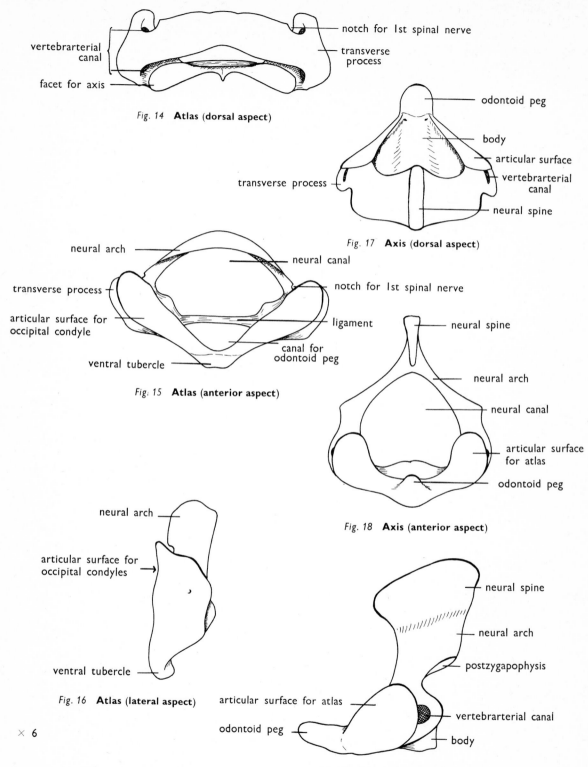

Fig. 14 **Atlas (dorsal aspect)**

notch for lst spinal nerve

transverse process

vertebrarterial canal

facet for axis

odontoid peg

body

articular surface

vertebrarterial canal

neural spine

transverse process

Fig. 17 **Axis (dorsal aspect)**

neural arch

neural canal

transverse process

notch for lst spinal nerve

articular surface for occipital condyle

ligament

ventral tubercle

canal for odontoid peg

Fig. 15 **Atlas (anterior aspect)**

neural spine

neural arch

neural canal

articular surface for atlas

odontoid peg

Fig. 18 **Axis (anterior aspect)**

neural arch

articular surface for occipital condyles

ventral tubercle

Fig. 16 **Atlas (lateral aspect)**

× 6

neural spine

neural arch

postzygapophysis

articular surface for atlas

vertebrarterial canal

odontoid peg

body

Fig. 19 **Axis (lateral aspect)**

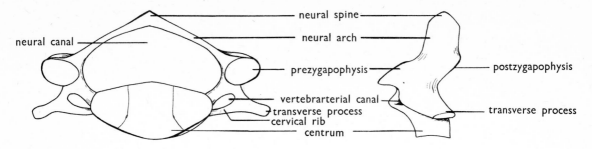

Fig. 20 **3rd cervical vertebra (anterior aspect)**

Fig. 21 **3rd cervical vertebra (lateral aspect)**

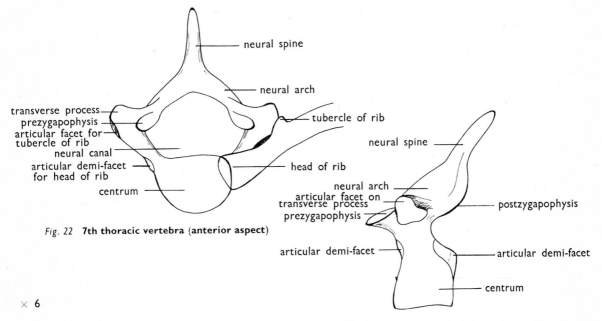

Fig. 22 **7th thoracic vertebra (anterior aspect)**

× 6

Fig. 23 **7th thoracic vertebra (lateral aspect)**

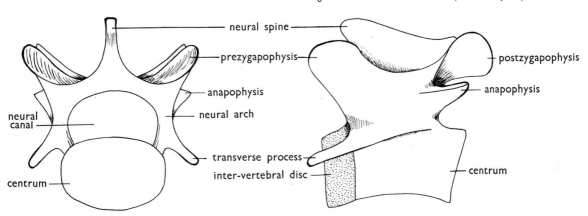

Fig. 24 **3rd lumbar vertebra (anterior aspect)**

Fig. 25 **3rd lumbar vertebra (lateral aspect)**

14

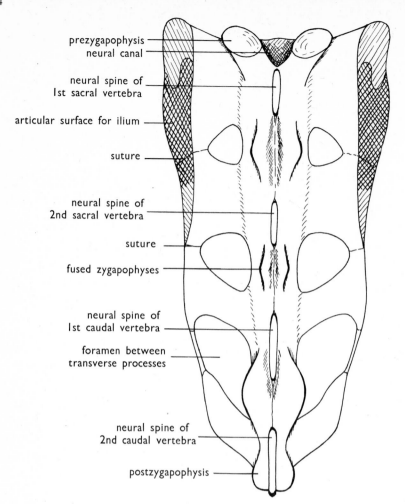

prezygapophysis
neural canal

neural spine of
1st sacral vertebra

articular surface for ilium

suture

neural spine of
2nd sacral vertebra

suture

fused zygapophyses

neural spine of
1st caudal vertebra

foramen between
transverse processes

neural spine of
2nd caudal vertebra

postzygapophysis

Fig. 26 **Sacrum (dorsal aspect)**

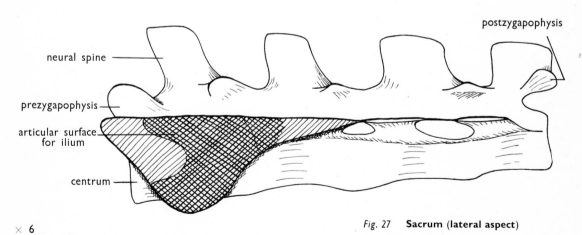

postzygapophysis

neural spine

prezygapophysis

articular surface
for ilium

centrum

× 6

Fig. 27 **Sacrum (lateral aspect)**

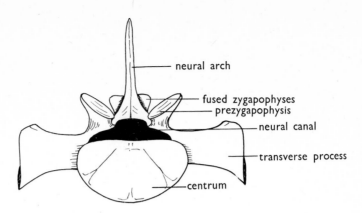

Fig. 28 **Sacrum (anterior aspect)**

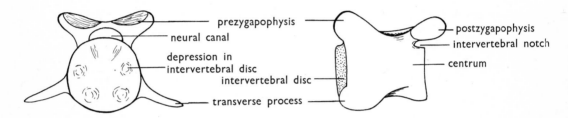

Fig. 29 **3rd free caudal vertebra (anterior aspect)**   Fig. 30 **3rd free caudal vertebra (lateral aspect)**

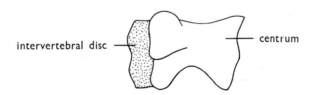

Fig. 31 **10th free caudal vertebra (lateral aspect)**

× 6

intervertebral disc —◼— centrum

Fig. 32 **24th free caudal vertebra (lateral aspect)**

(c) RIBS

The rat has thirteen pairs of ribs corresponding to the thirteen thoracic vertebrae.

Each rib has a long **shaft**, a **head** and a **tubercle** near the head. The heads of the ribs articulate with the bodies of the vertebrae and the tubercles articulate with the transverse processes. With the exception of the last two or three pairs of ribs, the heads overlap the intervertebral joints and thus articulate with demifacets on the vertebrae—see Fig. 22, p. 13.

The ventral ends of the 1st–11th pairs of ribs articulate with costal cartilages. These may become calcified late in life but never ossify. The costal cartilages of the 1st–7th pairs of ribs are attached directly to the sternum, those of the 8th ribs are attached to those of the 7th, the 9th to the 8th and the 10th to the 9th. The costal cartilages of the 11th ribs are free.

The ribs whose costal cartilages are attached to the sternum directly are called **true ribs**. The remaining ribs are called **false ribs**. Of the latter those without costal cartilages or whose costal cartilages are unattached, i.e. the 11th, 12th and 13th pairs of ribs, are called **floating ribs**.

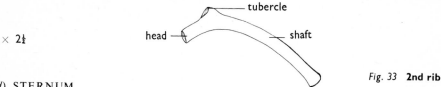

× 2½

Fig. 33 **2nd rib**

(d) STERNUM

The sternum is elongated and jointed. In the rat it consists of a **manubrium**, four **sternebrae**, and a **xiphisternum** to which a **xiphoid cartilage** is attached. The costal cartilages of the true ribs articulate with the manubrium and the sternebrae.

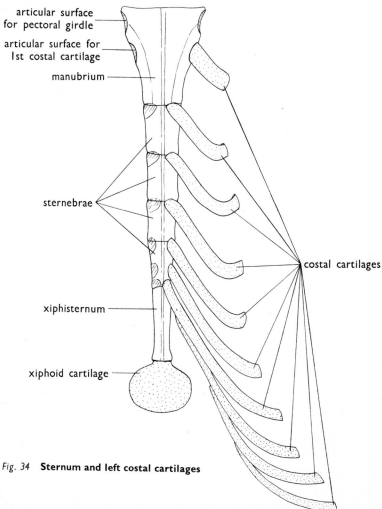

× 2½

Fig. 34 **Sternum and left costal cartilages**

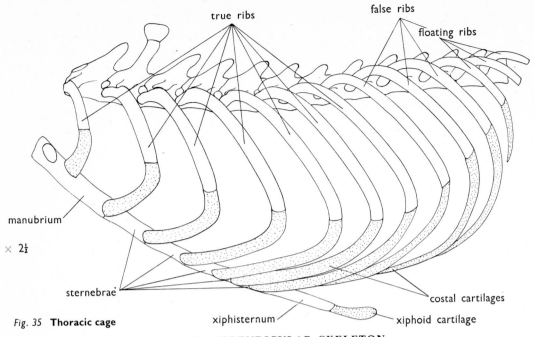

true ribs

false ribs

floating ribs

manubrium

$\times\ 2\frac{1}{2}$

sternebrae

xiphisternum

costal cartilages

xiphoid cartilage

Fig. 35 **Thoracic cage**

2. APPENDICULAR SKELETON

This consists of the **pectoral** and **pelvic girdles** and the bones of the **fore-** and **hind-limbs**.

(a) PECTORAL GIRDLE

The pectoral girdle consists of two **scapulae** and two **clavicles** with, in the rat, the vestiges of procoracoids between them and between the clavicles and the sternum. Parts of the procoracoids ossify as **omosterna** when the rat is about three months old.

The **scapulae** are cartilage bones and retain small unossified regions, the **suprascapular cartilages**. Each scapula has a flattened tapering blade with a conspicuous ridge called the **spine** which ends in an **acromion process** with a small **metacromion process** on its ventral border. The anterior end of the blade of the scapula is truncated to form the slightly concave **glenoid cavity** which is overhung by the **coracoid process**—a vestige of the coracoid bone of lower vertebrates. The procoracoid cartilages are similarly vestigeal structures because the procoracoid bones have been overlaid and replaced by the clavicle during the process of evolution.

The **clavicles** are membrane bones. Each clavicle is a slightly curved rod of bone. It is suspended between the acromion process of the scapula and the omosternum.

With the sternum, the pectoral girdle forms a complete arch so that the shoulders are strongly supported. Nevertheless they have the considerable freedom of movement necessary for burrowing.

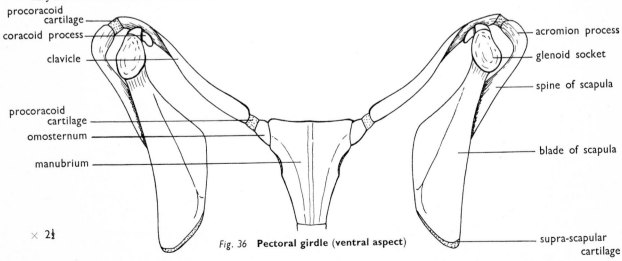

procoracoid cartilage

coracoid process

clavicle

procoracoid cartilage

omosternum

manubrium

acromion process

glenoid socket

spine of scapula

blade of scapula

supra-scapular cartilage

$\times\ 2\frac{1}{2}$

Fig. 36 **Pectoral girdle (ventral aspect)**

(b) PELVIC GIRDLE

The pelvic girdle is formed of two **innominate bones**. Each innominate bone is formed of an **ilium**, an **ischium** and a **pubis** fused together with, in the rat, a small **cotyloid bone**. The **ilia** are large with long wings which articulate with the sacrum at the **sacro-iliac joints**. The **pubes** meet one another at the **pubic symphysis**. Between each pubis and the corresponding ischium there is a large **obturator foramen**. The **cotyloid bone** lies between the other bones in such a way that the pubis does not take part in the formation of the **acetabulum**, the deep socket for the head of the femur.

With the sacrum, the pelvic girdle forms a complete ring and thus gives firm support to the hind limbs.

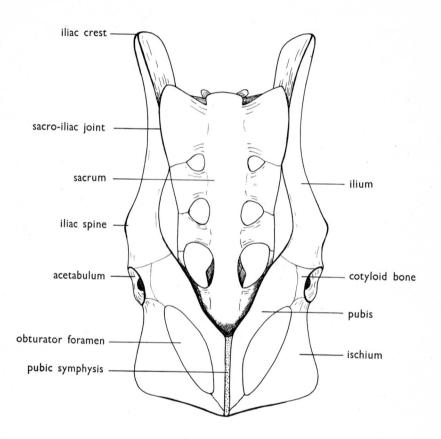

× 2½

Fig. 37 **Pelvic girdle (ventral aspect)**

(c) THE LIMBS

All vertebrates except the fishes possess limbs which are modified to a greater or lesser extent from the primitive **pentadactyl** pattern with five fingers and five toes. The modification in the rat is slight. In addition to the cartilage bones of the pentadactyl limb, there are a number of small membrane bones called **sesamoid bones** which develop in tendons in the regions of some of the joints.

The rat is **plantigrade**, i.e. both fore- and hind-feet are used flat on the ground.

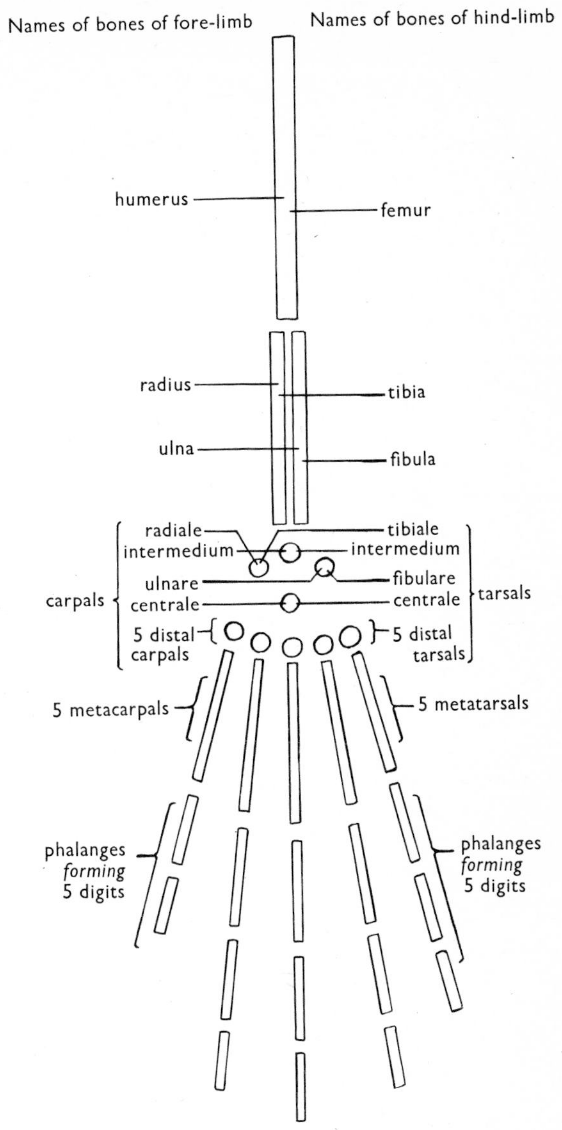

Fore-limb

The **humerus** has a long shaft with a tuberosity for the attachment of the deltoid muscle on its anterior side. The head of the humerus is rounded to fit the **glenoid cavity** and thus form the **shoulder joint**. Near the head of the humerus are the greater and lesser tubercles for the attachment of some of the shoulder muscles. The distal end of the humerus is grooved to form the **trochlea** which articulates with the ulna to form the **elbow joint**. Above the trochlea, at the back of the humerus, is the **olecranon notch** into which the **olecranon process** of the ulna fits when the forearm is extended. Above the front of the trochlea is the **supratrochlear notch** into which the **coronoid process** of the ulna and the **head** of the radius fit when the forearm is flexed. On either side of the trochlea there are ridges called **condyles**, the regions above which are the **epicondyles**.

The **radius** and **ulna** are not fused, but the radius is curved over the ulna so that the fore-foot is normally in the **prone** position (palm downwards) and cannot be fully supinated (turned palm upwards). The epiphyses on the distal ends of the radius and ulna remain distinguishable till very late in life.

The **carpals** are arranged in two rows. The proximal row consists of two bones, one of which is formed from the fusion of the **radiale** and the **intermedium** while the other is the **ulnare**. The distal row consists of five bones. The fourth and fifth **distal carpals** are fused but the **centrale** lies between the second and third distal carpals instead of lying between the rows.

Fig. 38 **Plan of pentadactyl limb**

The first **metacarpal** is very much shorter and thinner than the other four.

There are fourteen **phalanges**, two in the first digit and three in each of the other four digits. Those of the first digit are very short, so that the entire digit does not extend beyond the end of the second metacarpal. The terminal phalanx of each of the other digits is tapered to fit inside a strong **claw**.

There are three sesamoid bones in the wrist of the rat—the pisiform, the falciform and the ulnar sesamoid. There are also a pair of sesamoid bones at the distal ends of each of the metacarpals and a single sesamoid bone at the proximal end of each terminal phalanx.

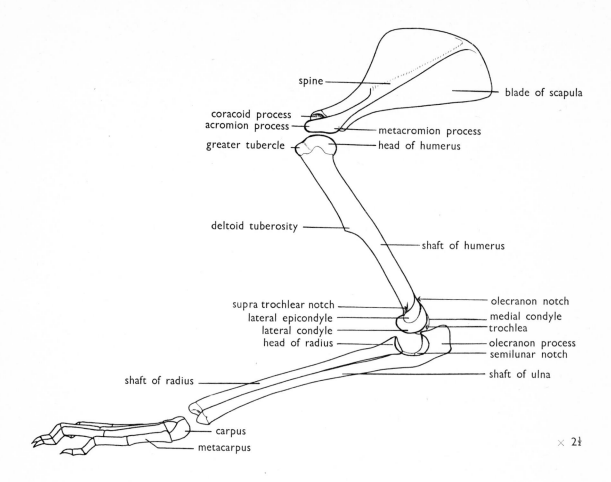

Fig. 39 **Left fore-limb and scapula, partially disarticulated (lateral aspect)**

Hind-limb

The **femur** has a long shaft. The **head** is rounded and is attached to the medial side of the proximal end of the shaft by a distinct neck. The head fits into the **acetabulum** to form the **hip joint**. Near the head there are three **trochanters** for the attachment of the hip muscles. The distal end of the femur has two rounded **condyles** on a conspicuous epiphysis.

The **tibia** and **fibula** are partially fused to form a **tibio-fibula**. The proximal ends of the bones are free and have distinct epiphyses. That of the tibia is grooved to fit the condyles of the femur. The distal region where fusion has taken place is grooved to fit the astragalus and has knobs on either side called **malleoli**.

The **tarsals** are arranged in two irregular rows with the **centrale** between them. The proximal row consists of three bones. The **fibulare** or **calcaneum** is very large and is prolonged under the **intermedium** or **astragalus** to provide attachment for the strong muscles of the calf of the leg. The astragalus has a cotton-reel-shaped surface for articulation with the tibio-fibula. The **tibiale** is small. The distal row consists of four bones because the fourth and fifth **distal tarsals** are fused.

The first **metatarsal** is somewhat shorter than the other four.

There are fourteen **phalanges**, two in the first digit and three in each of the other four digits. The terminal phalanx of each digit is tapered to fit inside a strong claw.

The **patella** or knee-cap is a large sesamoid bone over the knee joint. There are also a pair of sesamoid bones at the distal end of each metatarsal and a single sesamoid bone at the proximal end of each terminal phalanx.

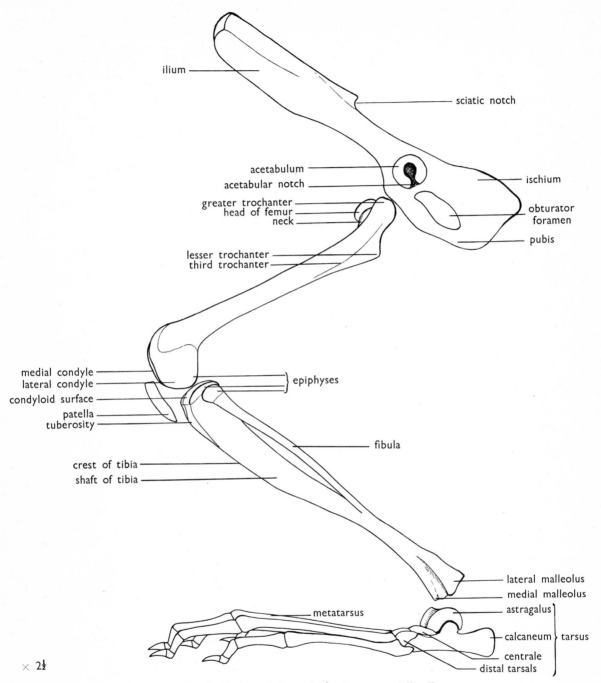

ilium

sciatic notch

acetabulum
acetabular notch

ischium

greater trochanter
head of femur
neck

obturator
foramen

pubis

lesser trochanter
third trochanter

medial condyle
lateral condyle
condyloid surface
patella
tuberosity

} epiphyses

fibula

crest of tibia
shaft of tibia

lateral malleolus
medial malleolus

metatarsus

astragalus
calcaneum } tarsus
centrale
distal tarsals

× 2½

Fig. 40 **Left hind-limb and innominate bone, partially disarticulated (lateral aspect)**

22

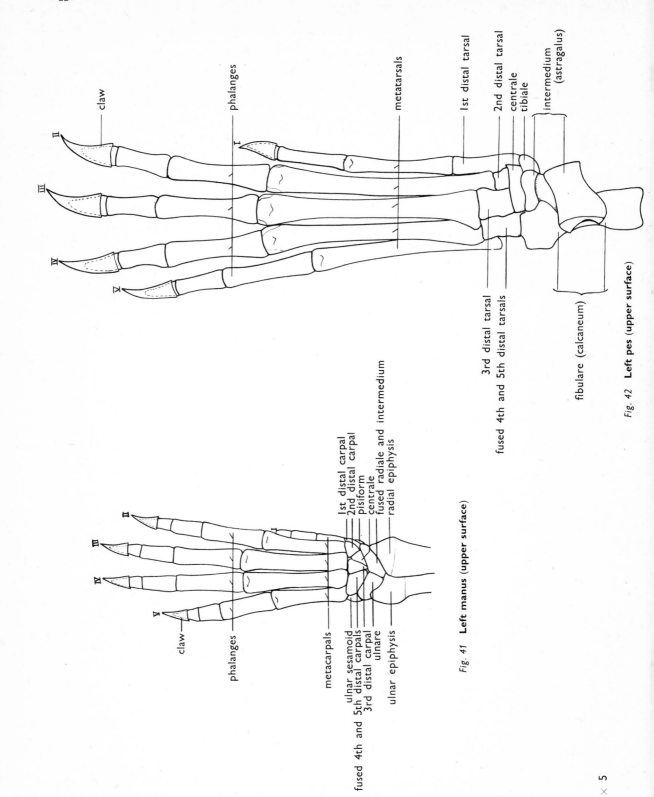

claw

phalanges

metatarsals

1st distal tarsal

2nd distal tarsal
centrale
tibiale

intermedium
(astragalus)

3rd distal tarsal

fused 4th and 5th distal tarsals

fibulare (calcaneum)

Fig. 42 **Left pes (upper surface)**

claw

phalanges

metacarpals

ulnar sesamoid
fused 4th and 5th distal carpals
3rd distal carpal
ulnare

ulnar epiphysis

1st distal carpal
2nd distal carpal
pisiform
centrale
fused radiale and intermedium
radial epiphysis

Fig. 41 **Left manus (upper surface)**

5

×

JOINTS

As in all bony vertebrates the joints are of three kinds.

(1) **Fixed** or **immovable joints**, e.g. the sutures of the skull. In such joints the bones are joined together by tough fibrous tissue which may ossify later in life.

(2) **Slightly movable joints**, e.g. the pubic symphysis and the majority of the joints between the bodies of the vertebrae. In such joints there is a pad of cartilage between the opposing bone surfaces and the bones are bound together by a capsule of tough fibrous tissue around the cartilage.

(3) **Freely movable** or **synovial joints**, e.g. the shoulder, hip, elbow and knee. The vast majority of the joints of the body are of this type. In such joints the articular surfaces are covered with cartilage and the bones are held together by a capsule of elastic fibrous tissue (ligament). The capsule is lined with **synovial membrane** which produces small quantities of **synovial fluid** to lubricate the movements between the opposing surface of the joint.

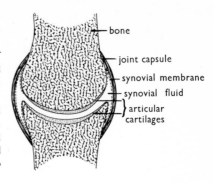

Fig. 43 **Structure of a typical synovial joint**

MUSCLES

THE muscular system is extremely complex. Except in the case of sphincter or ring muscles, each skeletal muscle has an **origin** and an **insertion**. The origin is the less movable or less frequently moved attachment of the muscle and the insertion is the more movable or more frequently moved attachment. The majority of muscles act in **antagonistic** groups, e.g. the **biceps brachii** flexes the elbow with the aid of the **brachialis** while the **triceps brachii** extends the elbow with some help from the **anconeus**.

Details of these muscles:

Muscle	Origin	Insertion
Biceps brachii	two heads (parts) from scapula	radius
Brachialis	humerus	ulna
Triceps brachii	three heads, one from scapula and two from humerus	ulna
Anconeus	humerus	ulna

In addition the **coracobrachialis** is shown in Fig. 44 because it originates with the short head of the **biceps**. It is inserted on the humerus and therefore does not assist flexion of the elbow.

The general plan of the muscular system of the rat is the same as that for other mammals, including man. Most of the muscles which are found in man have homologous muscles in the rat, though with slight differences, e.g.:

(1) the **biceps femoris** of the rat has actually three heads, not two, though its name is unchanged;

(2) the **levator auris** muscle of the rat is relatively much larger and stronger than the **auricularis** muscles of man which are vestigial and no longer able to "prick the ears";

(3) the much more extensive **platysma** of the rat takes the place of the numerous muscles of facial expression found in man;

(4) the very extensive **cutaneous** muscle of the rat extends over the trunk from the axillae to the base of the tail and has no homologue in man.

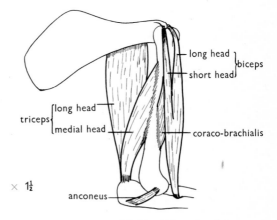

Fig. 44 **Medial view of the muscles of the upper fore-limb in the resting position**

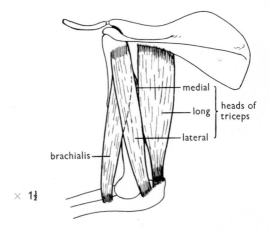

Fig. 45 **Lateral view of the muscles of the upper fore-limb in the resting position**

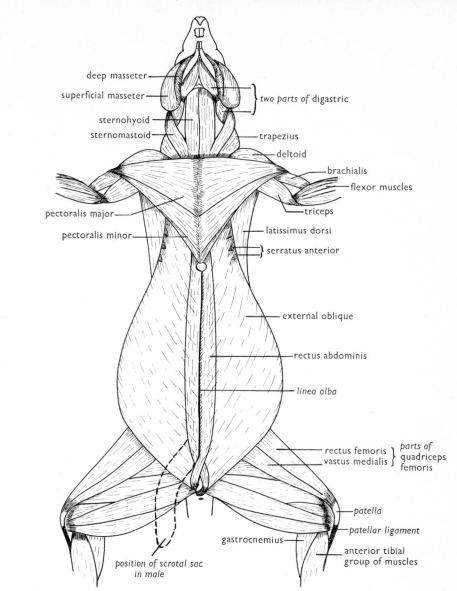

deep masseter

superficial masseter

} *two parts of digastric*

sternohyoid

sternomastoid

trapezius

deltoid

brachialis

flexor muscles

triceps

pectoralis major

pectoralis minor

latissimus dorsi

} serratus anterior

external oblique

rectus abdominis

linea alba

rectus femoris

vastus medialis

} *parts of* quadriceps femoris

patella

patellar ligament

gastrocnemius

anterior tibial group of muscles

position of scrotal sac in male

$\times \frac{2}{3}$

Fig. 46 Muscles (ventral view with platysma and cutaneous muscles and glands of the neck removed)

FUNCTIONING OF MUSCLE

Contraction of muscle may be **isotonic**, i.e. shortening without increase in tension, or **isometric**, i.e. increasing in tension without shortening, or a combination of both. The strength of contraction of the muscle as a whole depends on the proportion of the fibres which are stimulated and contract. **Tone** of a muscle is produced by fibres contracting in turn.

During muscular contraction a large number of chemical changes take place. Energy for contraction is provided by the breakdown of **adenosinetriphosphate (ATP)** to **adenosinediphosphate (ADP)**. The ATP is resynthesized from ADP plus activated phosphate ($\backsim$ P). The energy required for this activation is derived from the organic nutrients of food. **Glucose** and **glycogen** undergo **glycolysis** to **ketopyruvic acid** yielding 2 and 3 moles ATP per hexose unit respectively. This process is **anaerobic**. If oxygen is not available the pyruvic acid is converted to lactic acid which accumulates and interferes with the activity of the enzymes to such an extent that no further energy is made available and contraction of muscle ceases. If oxygen is available the pyruvic acid undergoes **oxidative decarboxylation** via the **tricarboxylic acid cycle** which yields **carbon dioxide** and high-energy **hydrogen** atoms which join those derived from glycolysis. In the **oxidative phosphorylation** that follows, the hydrogen atoms are not directly involved but the energized **electrons** from them pass down the **electron transfer** chain of **cytochromes**, giving up energy to produce an average of 3 moles ATP per pair of electrons (36 moles ATP per hexose unit) and finally causing oxygen to combine with free hydrogen ions to form **water**. The tricarboxylic cycle can make use of substrates derived from fats and proteins as well as from carbohydrates, with an overall energy conversion efficiency in the cell of just under 50%. Lactic acid or pyruvic acid escaping into the blood stream are resynthesized to glycogen in the liver by energy derived from the oxidative phosphorylation there.

Details of the physiology of muscle action are outside the scope of this book. When muscle contracts **isometrically** all the energy eventually appears as heat. When it contracts **isotonically** 20 to 25% of the energy appears as mechanical work. The average efficiency of muscle is about 15%. If activity continues long enough to use up all the reserves of glycogen in the muscle, no further activity can take place until they have been replenished. The muscle is **fatigued** and requires a period of **rest**.

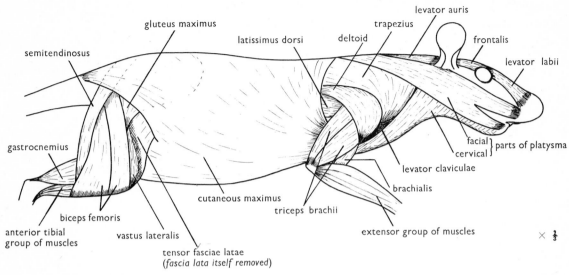

Fig. 47 **Chief superficial muscles (lateral view with fascial sheath removed from head and limbs)**

BODY CAVITIES

MAMMALS are **coelomate**, but the coelom is restricted to the trunk region. It is divided by a muscular partition, the **diaphragm**, into a **thoracic cavity** and an **abdominal cavity**. The thoracic cavity is subdivided by non-muscular membranes so that the heart and the lungs lie in separate pouches known as the **pericardial** and **pleural cavities** respectively.

The entire coelom is lined with serous membrane which also covers all the organs in the cavities. This membrane forms the **pericardium** round the heart, the **pleurae** round the lungs and the **peritoneum** lining the abdomen and covering the abdominal viscera. The pericardium and pleurae form distinct double-walled bags because the organs they cover are more or less the same shape as the spaces in which they lie, but the peritoneum forms a sheath round a number of organs of various shapes and also forms suspensory folds, which are called **mesenteries** where they support parts of the alimentary canal, and **mesovaria** where they support the ovaries and oviducts of the female.

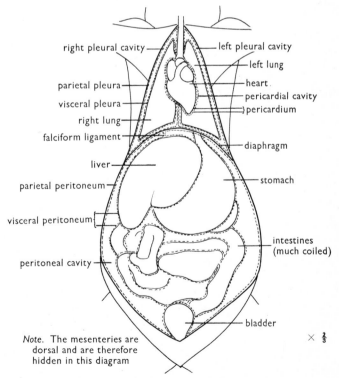

Note. The mesenteries are dorsal and are therefore hidden in this diagram

Fig. 48 **Body cavities, showing the relationship of the serous membranes to the viscera**

DIGESTIVE SYSTEM

THE digestive system of any mammal consists of the **alimentary canal** and its **associated glands**. The canal starts at the **mouth** and ends at the **anus**. It is divided into the following regions:

buccal cavity, pharynx, oesophagus, stomach, small intestine, large intestine.

The associated glands are the **salivary glands**, the **pancreas** and the **liver**.

Digestion is the means whereby the food is rendered suitable for absorption into the tissues.

(1) **Mechanical digestion** is the breakdown of large into smaller particles by purely mechanical means.

(2) **Chemical digestion** is the breakdown of the large molecules in the food into smaller molecules which can be more readily absorbed. It chiefly concerns the three main categories of food materials—carbohydrates, fats and proteins.

(i) **Carbohydrates.** Monosaccharides (glucose and fructose) require no digestion but the polysaccharides (e.g. starch) and the disaccharides (maltose, sucrose and lactose) are broken down into their component monosaccharides.

(ii) **Fats.** Some fats are absorbed without digestion but others are broken down into fatty acids and glycerol.

(iii) **Proteins.** Proteins are long chains of amino acids joined by links between the amino and the organic acid groups. The twenty-two amino acids which occur in nature are arranged in an infinite number of ways to form the different proteins. Digestion involves breaking the long chains into shorter chains called proteoses and peptones and breaking off the ends of the chains until the amino acids are freed from one another.

In all cases the breakdown of the large molecules of the food substances is due to **hydrolysis**, i.e. a water molecule is used to satisfy the broken bond. The hydrolysis is catalysed by **enzymes** which are highly specific, i.e. a special enzyme is needed for each type of substance (substrate) and each will only act in a limited range of conditions. The enzymes are themselves proteins and are ultimately destroyed.

BUCCAL CAVITY

The buccal cavity is the region between the mouth and the pharynx. It is supported by the **jaws**, both of which bear **teeth**. It has the **cheeks** for its sides, the **tongue** on its floor and the **palate** as its roof.

The **jaws** of the rat are long and narrow. The upper jaw is fixed but the lower jaw is movably articulated to the rest of the skull. The joint lies just in front of the ear. It is condyloid in form, and in the rat it is very shallow, so that rotary as well as opening and closing movements can be made. The jaw is moved by strong muscles in the cheeks, the most important of which are the **masseter**, **temporalis** and **pterygoid** muscles.

TEETH

The teeth of mammals are set in sockets in the jaw-bones. The distal or exposed part of each tooth is the **crown**, while the part within the jaw-bone forms one or more **roots**.

As in the teeth of other vertebrates, the bulk of the tooth substance is **dentine**. The crown is covered partially or entirely with **enamel**, the hardest substance in the body, and the roots are often invested in **cement**. All these substances are heavily calcified but differ in origin and arrangement of the deposits. Enamel is formed by calcification of parts of special cells called **ameloblasts**, while dentine is the calcified matrix produced by cells called **odontoblasts** and cement is structurally very like bone.

The core of each tooth is the pulp cavity filled with **dental pulp**, a soft tissue containing **blood-vessels** and **nerves**.

Mammals are **heterodont**, i.e. their teeth are not all alike but are divided into groups, the **incisors** in front, then the **canines** and then the **premolars** and **molars** at the back. The incisors and canines are single-rooted while the premolars and molars often have two or more roots. The crowns of the incisors and canines are usually sharp-edged or pointed with single cusps, while those of the premolars and molars have many cusps and are adapted to the type of food, e.g. sharp-edged for cutting meat in carnivores and flattened for grinding in herbivores.

Most mammals are **diphyodont**, i.e. they have two sets of teeth. The first set or milk dentition consists of incisors, canines and milk molars. These are replaced by the adult incisors, canines and premolars, while the adult molars are added as the jaw enlarges sufficiently to accommodate them. In any species of mammal the numbers of the different types of teeth may be expressed by the dental formula. The primitive maximum dentition is:

$$\text{I } \frac{3}{3}; \text{C } \frac{1}{1}; \text{PM+M } \frac{7}{7}=44.$$

The teeth of the rat are very highly specialized. The incisors grow continuously throughout life. Neither they nor the molars are replaced and there are no canines. Thus the rat is **monophyodont** with one dentition only. The dental formula is:

$$\text{I } \frac{1}{1}; \text{C } \frac{0}{0}; \text{PM } \frac{0}{0}; \text{M } \frac{3}{3}=16.$$

The gap between the incisors and the molars is known as the **diastema**.

The **incisors** are present in the jaw at birth and erupt into the buccal cavity on about the tenth day. Each incisor has a powerful crown and no root. It is roughly triangular in transverse section and is a segment of a flat spiral. The upper incisors are anterior to the lower incisors and are segments of a smaller spiral, i.e. they have greater curvature.

The labial or anterior surface of each incisor is covered with enamel which is formed from a persistent enamel organ and

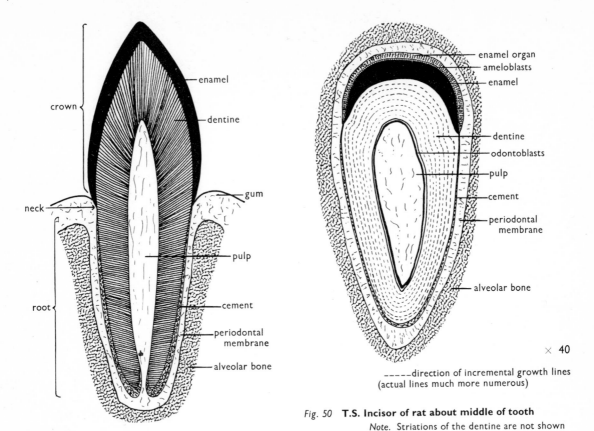

Fig. 49 **V.S. Typical mammalian incisor**

Fig. 50 **T.S. Incisor of rat about middle of tooth**

Note. Striations of the dentine are not shown

- - - - - direction of incremental growth lines
(actual lines much more numerous)

× 40

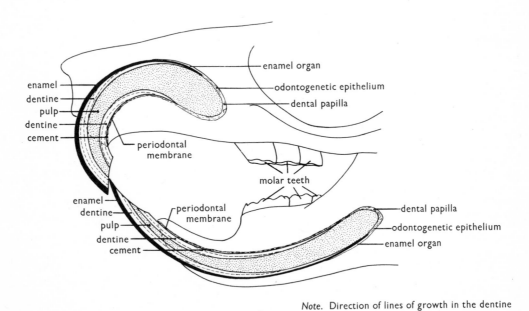

× 3

Note. Direction of lines of growth in the dentine

Fig. 51 **Parts of jaws with V.S. incisor teeth**

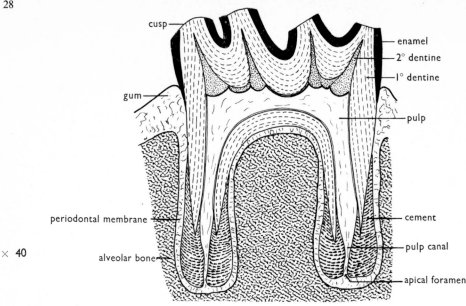

cusp

enamel
2° dentine
1° dentine

gum

pulp

periodontal membrane

cement

× 40

pulp canal

alveolar bone

apical foramen

- - - - - direction of incremental growth lines
(actual lines much more numerous)

Note. Striations of dentine are not shown

Fig. 52 **V.S. First molar of rat**

is pigmented bright orange from the fifth week onwards. The lingual or posterior and the lateral surfaces are covered with a very thin layer of cement. As enamel is harder than dentine and cement, there is differential wear which produces the chisel edge characteristic of the incisors of all rodents. Normally, wear of the incisors keeps pace with growth, but if one tooth is damaged the one opposite may overgrow and eventually perforate the palate or lip, causing death by starvation. The upper incisors of the rat grow 2·1 mm per week and the lower incisors 2·8 mm per week.

The enamel and dentine show daily incremental rings due to rhythmic calcification which gives rise to dense and less dense layers. The appearance of the layers varies with health, growth and nutrition. A record of any change is left until the affected part of the tooth is eventually worn away about forty-six days later.

The **molars** are molariform, i.e. each molar has several cusps the tips of which are free from enamel. Each molar also has several roots, all of which possess cement which increases in quantity with age.

The molars erupt on the nineteenth, twenty-third and thirty-fifth days respectively. The rate of eruption is rapid until contact with the appositional tooth is made. Thereafter it is much slower and the enamel organ degenerates so that further formation of enamel is impossible. Growth of primary dentine continues till the 125th day. After this, eruption to compensate for wear is due entirely to the formation of cement, so that in the adult one-third or more of the root may be of this substance. At the same time attrition stimulates the formation of secondary dentine in the regions of the pulp cavity which lie under the cusps.

The lamellations of the dentine record changes in the health and nutrition of the individual during the first 125 days of life. The record is permanent, lasting throughout life.

TONGUE

The **tongue** is very muscular and, in the rat, has a ridged and roughened surface. The muscles are of two types: (*a*) **extrinsic muscles** which originate from bones of the skull and move the whole tongue, and (*b*) **intrinsic muscles** which lie entirely within the tongue and alter its shape. In the resting position the ridges of the tongue of the rat fit into grooves on the palate so that the buccal cavity is virtually occluded. This prevents inhalation of dust while gnawing. The tongue is used to move the food about in the buccal cavity, to press it between the molar teeth so that it can be chewed, and to mix it with the saliva from the salivary glands. On the surface of the tongue are **taste buds** sensitive to certain types of stimuli, but the majority of flavours are in fact smelt.

PALATE

The **palate** separates the buccal cavity from the nasal cavities. The front part of the palate is supported by bone and is known as the **hard palate**, while the back part is of membrane only and forms the **soft palate**. The edge of the soft palate acts as a **valve** to prevent food from going up the nose during swallowing. Like the tongue, the palate has a few taste buds.

PHARYNX

The pharynx is the region between the buccal cavity and the oesophagus and also between the nasal cavities and the windpipe. It is thus shared by the digestive and the respiratory systems. Normally, food only passes through the posterior

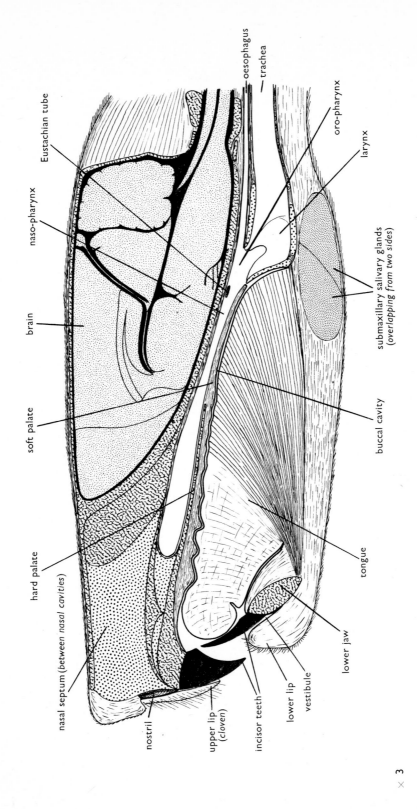

Fig. 53 **Median vertical section through the head showing the relationship of the buccal cavity, pharynx and oesophagus to the respiratory passages**

oesophagus

trachea

oro-pharynx

larynx

Eustachian tube

naso-pharynx

submaxillary salivary glands
(overlapping from two sides)

brain

soft palate

buccal cavity

hard palate

tongue

nasal septum *(between nasal cavities)*

lower jaw

nostril

vestibule

upper lip *(cloven)*

lower lip

incisor teeth

3

×

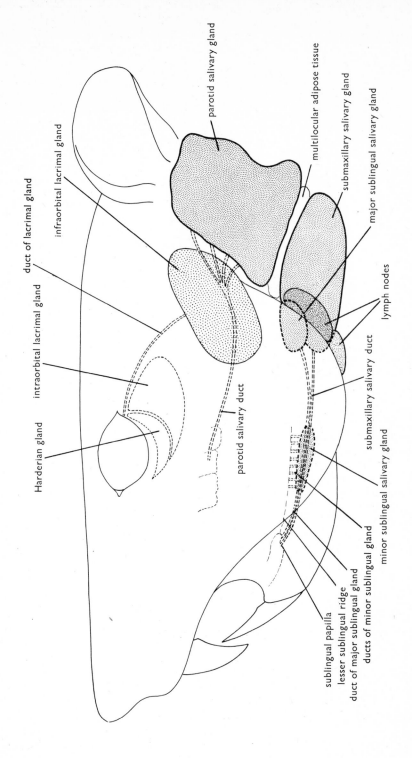

Fig. 54 **Salivary glands and associated structures**

× 3

parotid salivary gland

multilocular adipose tissue

submaxillary salivary gland

major sublingual salivary gland

lymph nodes

infraorbital lacrimal gland

duct of lacrimal gland

intraorbital lacrimal gland

Harderian gland

parotid salivary duct

submaxillary salivary duct

submaxillary salivary gland

minor sublingual salivary gland

sublingual papilla

lesser sublingual ridge

duct of major sublingual gland

ducts of minor sublingual gland

part of the pharynx, which is known as the **oro-pharynx**, while air passes through both the naso-pharynx and the oro-pharynx.

The opening into the windpipe lies on the ventral side of the pharynx and is known as the **glottis**. This opening can be closed and covered by the **epiglottis** so that food cannot pass through it during swallowing. In the rat the epiglottis extends to the edge of the soft palate.

SALIVARY GLANDS

The salivary glands are associated with the buccal cavity into which their ducts open. The rat has four pairs of these glands—parotid, submaxillary, major and minor sublingual.

(1) **Parotid**—large and irregular; behind and below the ears with ducts opening opposite the upper molars.

(2) **Submaxillary**—large and ovoid; under the chin and extending down the neck almost to the sternum, with ducts opening on the sublingual folds under the front of the tongue.

(3) **Major sublingual**—small and rounded; close in front of the submaxillary glands, with ducts opening into those of the latter.

(4) **Minor sublingual**—flattened and irregular; under the sides of the tongue, with ducts opening on the lesser sublingual ridges lateral to the sublingual folds.

Note. The human has only three pairs of salivary glands, the ducts of which show them to be homologous with the parotid, submaxillary and minor sublingual glands of the rat.

The salivary glands produce **saliva**. This is a mixed secretion consisting of **mucus** and a watery solution of the enzyme **salivary amylase** (ptyalin). It softens the food and makes it slippery for passage through the pharynx and oesophagus while the enzyme starts the chemical digestion of **starch**. The action on uncooked starch is slow, but when the envelopes of the starch grains are ruptured by cooking the action is extremely rapid.

OESOPHAGUS

The oesophagus is a straight tube from the pharynx to the stomach. It lies partly in the neck and partly in the thorax, dorsal to the heart and between the lungs. It passes through a perforation in the diaphragm before reaching the stomach; therefore there is a very short section of it in the abdomen. The wall of the oesophagus has longitudinal and circular layers of muscle which, acting antagonistically, produce characteristic undulating or **peristaltic movements** to assist the passage of the food mass or **bolus**. At rest the oesophagus is dorso-ventrally flattened so that its lumen is occluded and it appears very narrow in Fig. 53.

STOMACH

The stomach is large and saccular. It lies to the left close below the diaphragm, and in the rat is divided into two distinct regions. The left region is whitish and translucent. It has numerous **mucous glands** but no digestive glands. The right region is reddish-grey, highly vascular and muscular and appears opaque. It has numerous internal longitudinal folds and numerous digestive glands known as **peptic glands** because they produce the precursor of the enzyme **pepsin**. They also produce the precursor of the enzyme **rennin**. These glands are stimulated to activity by the hormone **gastrin**, which is produced by the stomach wall in response to the presence in the food of substances called **secretagogues**.

The stomach content of the rat has a pH of 3·6, though no free acid has been found in it. This acidity stops the action of salivary amylase, but allows **pepsin** to start the hydrolysis of **proteins**, splitting some of them into **proteoses** and **peptones**. It can attack the bonds between certain types of amino acids. **Rennin** curdles **milk** by converting the soluble milk protein **caseinogen** only into insoluble **casein**. It is important in the young before weaning because at this stage the stomach content is not fully acid and peptic digestion is impossible. Later the acidity of the gastric juice curdles any milk protein taken without the intervention of rennin, though the enzyme continues to be secreted in small quantities.

The opening from the oesophagus to the stomach is guarded by the **cardiac sphincter** muscle which prevents regurgitation of the food into the oesophagus and pharynx. The opening from the stomach into the intestine is guarded by the **pyloric sphincter** muscle which controls the flow of semi-digested food (**chyme**) into the small intestine.

SMALL INTESTINE

The small intestine of the rat is about six times the length of the body from snout to anus. This relatively enormous length is necessary to hold the food long enough for thorough digestion and to provide the large surface area needed for absorption. The relative length is even **greater** in **herbivores** such as the rabbit, in which the plant food is more slowly attacked by the digestive enzymes, but is **less** in **carnivores** such as the cat, in which the natural food, flesh, is more rapidly digested.

The small intestine is approximately uniform in diameter and much coiled to fit into the abdominal cavity. It is supported by an extensive mesentery so that the position of the coils can be altered according to the pressure from the other abdominal viscera and the abdominal wall.

The first loop of the intestine is called the **duodenum**. The **bile duct** from the **liver** opens into this region. In the rat this duct is joined by numerous **pancreatic ducts** from the **pancreas**, in man by a single main pancreatic duct, while in the rabbit the pancreatic duct opens into the duodenum separately. In man the remainder of the intestine is divided into **jejunum** and **ileum**, but in the rat it is all known as the **ileum**.

The wall of the small intestine is thin, but it contains bands of circular and longitudinal muscle which produce **peristaltic** and **antiperistaltic movements** responsible for rocking the contents to and fro, mixing it and gradually passing it towards the large intestine.

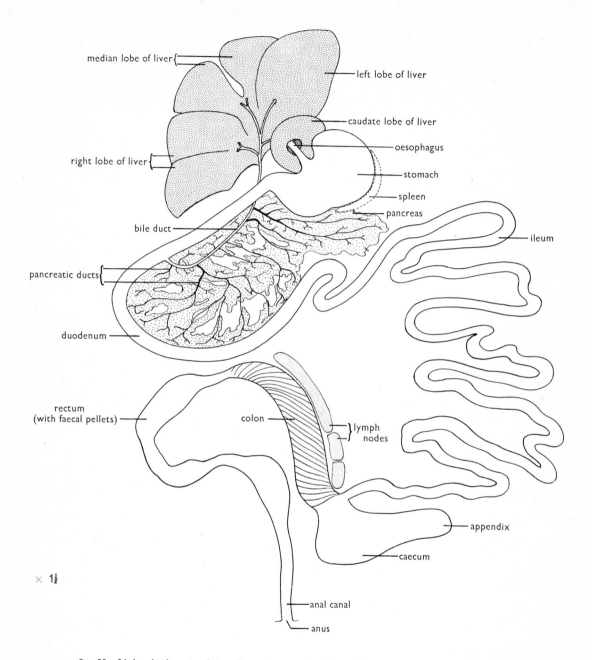

median lobe of liver

left lobe of liver

caudate lobe of liver

oesophagus

right lobe of liver

stomach

spleen

pancreas

bile duct

ileum

pancreatic ducts

duodenum

rectum
(with faecal pellets)

colon

lymph
nodes

appendix

caecum

× 1½

anal canal

anus

Fig. 55 **Abdominal parts of the alimentary canal and associated structures (displayed)**

The lining of the small intestine is thrown into **transverse folds** which are covered by numerous minute projections called **villi**. Each villus contains a network of fine **blood capillaries** around a lymph vessel called a **lacteal**. Between the villi are tubular **intestinal glands**. The mucous membrane of the villi consists mainly of **absorptive cells** whose free surfaces have **striated borders**, but there are also **goblet cells** which secrete mucus.

Hormones are produced by the intestine in response to the presence of food. **Secretin** and **pancreozymin** stimulate production of **pancreatic juice**, while **enterocrinin** stimulates production of **intestinal juice**. Pancreatic juice and intestinal juice together with bile complete the digestion of food. Pancreatic juice and bile are both alkaline and together neutralize the acidity of chyme so that the pH is correct for the activity of the pancreatic and intestinal enzymes.

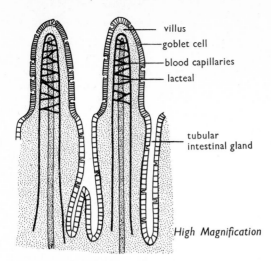

villus
goblet cell
blood capillaries
lacteal

tubular
intestinal gland

High Magnification

Fig. 56 **Portion of the lining of the small intestine (highly diagrammatic)**

Pancreatic juice contains pancreatic lipase, pancreatic amylase and the precursors of trypsin and chymotrypsin. **Pancreatic lipase** hydrolyses some **fats** to **fatty acids** and **glycerol**. **Pancreatic amylase** hydrolyses **starch** and **dextrin** to **maltose**. **Trypsin** and **chymotrypsin**, in an alkaline medium, act on **proteins, proteoses** and **peptones**. These enzymes attack the bonds adjacent to different amino acids. The combination of this action with the previous action of pepsin in the stomach results in the liberation of some **amino acids**, but the greater part of the protein is still incompletely digested, i.e. still **polypeptide**.

Intestinal juice, produced chiefly by the glands of the duodenum, but with contributions from shed epithelial cells, intestinal bacteria and mucus from goblet cells, contains the enzymes enterokinase, peptidases, maltase, sucrase and lactase. **Enterokinase** hydrolyses the precursor of trypsin, **trypsinogen**, to **trypsin**. Trypsin assists this action and also activates **chymotrypsinogen** to **chymotrypsin**. Thus neither of the protein-splitting enzymes becomes active till it reaches the intestine and the enterokinase. There are three intestinal **peptidases**, two of which can hydrolyse the terminal bonds of the acid and alkaline ends of the polypeptide

molecules respectively (**carboxypeptidase** and **aminopeptidase**). The third (**dipeptidase**) attacks the final bond between any two amino acids of a **dipeptide**, thus completing the digestion to **amino acids**.

Maltase, sucrase and **lactase** hydrolyse the corresponding disaccharides, **maltose** (from starch), **sucrose** (cane sugar) and **lactose** (milk sugar) into their constituent **monosaccharides**—mainly **glucose** with some **fructose** and **galactose**. The final products of protein and carbohydrate digestion, i.e. the amino acids and monosaccharides, are absorbed through the walls of the villi and pass directly into the blood stream.

Bile is poured on to the chyme as it enters the duodenum from the stomach. It is alkaline by nature of the bile salts though it contains no free alkali. Its digestive activity does not depend on enzymes. **Bile salts** assist the **emulsification of fats**, thus increasing the surface area for the action of lipase and also helping to keep the neutral fats in a state in which they can be absorbed directly. The **fatty acids** produced during the digestion of certain fats are absorbed in combination with the bile salts, which are thus conserved for future use. While the fatty acids pass directly into the blood stream, the emulsified neutral fats pass into the **lacteals** giving the lymph in them a milky appearance. This fat-bearing lymph is called **chyle**. It passes through the lymphatics of the mesenteries to a chain of large lymph nodes and thence through the thoracic duct to the blood stream.

LIVER

The liver is the largest gland in the body. It is suspended from the posterior surface of the diaphragm by the **falciform ligament**. In the rat it has four lobes. The **right lobe** is large and partially subdivided, the **left lobe** is also large but undivided, the median or **cystic lobe** has a deep fissure for the hepatic ligament and the **caudate lobe** is a small portion of liver tissue wrapped round the oesophagus.

The **bile duct** receives tributaries from each of the lobes and has a muscular sphincter at its duodenal end, but the rat has no gall bladder. When a **gall bladder** is present, e.g. in man and the mouse, it serves to store the bile secreted by the liver and to concentrate it by reabsorption of some of its water. Such storage and concentration do not occur in the rat, though the sphincter controls the delivery of bile into the duodenum.

The functions of the liver are numerous.

(1) It secretes **bile** which contains bile salts and bile pigments. As described above the **bile salts** aid the digestion and absorption of fats. They also aid the absorption of vitamins D, E and K and of carotene though not of vitamin A itself. They have antiputrifactive and laxative effects which are attributed mainly to the efficient digestion and absorption of fat in their presence. As they are absorbed with the fatty acids they are used over and over again. The **bile pigments** are waste materials from the breakdown of worn-out red blood corpuscles and are therefore excretory products. They colour the faeces.

(2) It stores carbohydrate in the form of **glycogen** and also stores **vitamins A, B** and **D** and **iron**.

(3) It affects metabolism by **deaminating** excess and unwanted **amino acids**, converting their nitrogenous parts into **urea**, which is later excreted by the kidneys, and converting their organic parts into carbohydrates or fats which are then utilized as such. The liver also **desaturates fats**, i.e. prepares them for the oxidation processes by which they are utilized.

PANCREAS

The pancreas of the rat is described as dendritic because it is very diffuse with small lobules scattered in the mesentery of the duodenal loop and in the fold called the gastro-splenic omentum. It has many ducts opening into the bile duct.

The pancreas produces **pancreatic juice**, the function of which is described above. It also possesses groups of cells called **islets of Langerhans** which produce the hormones **insulin** and **glucagon** concerned with the uptake and transfer of carbohydrate and therefore necessary for its correct utilization—see page 62.

LARGE INTESTINE

The large intestine of the rat is about one-sixth the length of the small intestine. Like the latter it is coiled and supported by mesentery. It is divided into a short **colon** and a longer **rectum**. The lining of the colon has conspicuous diagonal folds which are visible through the wall. At the junction of the colon and the ileum there is a diverticulum called the **caecum**, the blind end of which forms the **appendix**. The caecum of the rat has no internal septa such as are present in many rodents, but there is an **ileo-caecal** valve at its mouth. The appendix is marked off by a constriction and by the presence of a mass of lymphoidal tissue in its lateral wall.

By the time food reaches the large intestine, digestion is virtually complete and most of the products have been absorbed. The residues are partially dried by the **absorption** of **water** from them during their passage through the large intestine, and they become the semi-solid masses known as **faeces**. Passage of the faeces is lubricated by **mucus** from numerous unicellular glands called **goblet cells**. The faeces are voided periodically through the anus, the process of defaecation being under voluntary control so that the rat does not normally foul its sleeping quarters. The rat is, however, a great scavenger and to a certain extent **coprophagous**. This eating of the faeces is the usual method of infection with the numerous types of intestinal parasite which are common among rats.

In addition to any possible parasites, the large intestines of the majority of mammals contain a number of kinds of *Bacteria* and *Protozoa* which are **non-pathogenic**. These organisms not only do no harm but in certain cases perform the useful function of breaking down otherwise indigestible substances such as **cellulose**, and are therefore **symbiotic**. Symbiotic bacteria, found mainly in the vast caecum, are of great importance to rabbits, which eat relatively enormous amounts of green vegetation containing large quantities of cellulose. Such symbiosis is negligible in man and probably of little importance in the rat.

SUMMARY OF DIGESTION

Mechanical digestion takes place in the mouth through the **grinding** action of the molar teeth and also in the stomach through the **churning** action of the strong muscles of the stomach wall.

Mucus, secreted by the salivary glands, by the gastric glands of the stomach and by the unicellular glands (goblet cells) of the intestines, makes the food slippery and thus helps to prevent damage of the lining of the alimentary canal by rough particles.

Chemical digestion takes place throughout the alimentary canal, though very little occurs after the food reaches the large intestine. Digestion of different types of food substances is brought about by specific **enzymes** in different regions of the canal.

(i) Carbohydrate digestion

Starch can be hydrolysed to **dextrin** and **maltose** in the mouth and during passage to the stomach by the action of salivary amylase from the **saliva**. The action on cooked starch is extremely rapid, so that most of such starch is digested before the acidity of the gastric juice inactivates the enzyme. The action on raw starch is, however, very slow and most of such starch passes through the stomach to be hydrolysed in the intestine by pancreatic amylase from the **pancreatic juice**. Pancreatic amylase also completes the hydrolysis of any dextrin formed.

Maltose is hydrolysed to **glucose** in the intestine by **maltase** from the **intestinal juice**.

Sucrose or cane sugar is hydrolysed to **glucose** and **fructose** in the intestine by **sucrase** from the **intestinal juice**.

Lactose or milk sugar is hydrolysed to **glucose** and **galactose** in the intestine by **lactase** from the **intestinal juice**.

There are no cellulose-splitting enzymes in the mammal, but bacteria which utilize cellulose for their nourishment produce small quantities of glucose and fatty acids as waste products which are then available to the mammal.

(ii) Protein digestion

Proteins are hydrolysed by **peptidases**. **Endopeptidases** (pepsin, trypsin and chymotrypsin) attack the bonds adjacent to specific amino acids wherever they occur in the protein molecules. **Exopeptidases** (carboxy-peptidase, amino-peptidase and dipeptidase) can only attack the terminal bonds of the chains. Thus the action of the individual peptidases on the protein of food as it passes through the alimentary canal is as follows:

Pepsin, secreted as **pepsinogen** in the **gastric juice** and activated by itself in the stomach, starts the hydrolysis of certain **proteins** to **proteoses** and **peptones**. It is effective only in an acid medium, therefore its action stops as soon as the chyme is mixed with bile in the duodenum and the acidity is neutralised.

Trypsin, secreted as **trypsinogen** and **chymotrypsin** secreted as **chymotrypsin** in the **pancreatic juice** are rendered active in the small intestine. The activation starts by the action of **enterokinase**, in the intestinal juice, on some of the trypsinogen. Thereafter the trypsin itself activates more trypsinogen and the chymotrypsinogen. In the intestine these powerful protein-splitting enzymes are considerably diluted and the living cells can resist their action though the dead cells of food are destroyed. Together trypsin and chymotrypsin in an alkaline medium hydrolyse **proteins** remaining after the action of pepsin into **proteoses** and **peptones** and some **amino acids**.

Carboxy-peptidase, amino-peptidase and **dipeptidase** from the intestinal juice complete the hydrolysis of **proteoses** and **peptones** to **amino acids**.

(iii) **Fat digestion**

    Fats can be hydrolysed to **fatty acids** and **glycerol** by **lipases**. **Pancreatic lipase** from the pancreas acts in the intestine.

    Bile salts increase the activity of lipase by **emulsifying** fats so that the surface area exposed to enzyme action is increased, and some fats can be absorbed directly without hydrolysis. They are also essential for the absorption of fatty acids.

    Note. In man a certain amount of lipase is present in the stomach, though whether it is secreted by the stomach itself or is regurgitated from the duodenum is a debatable point. This **gastric lipase** starts the digestion of fats, but its action is relatively slight.

Enzyme	Precursor	Where produced	Where active	Activated by	Effect
Salivary amylase (ptyalin)		Salivary glands	Mouth and 1st part of stomach		Starch ⟶ dextrin and maltose
Pepsin	Pepsinogen	Peptic glands of stomach wall	Stomach	Pepsin already in stomach	Proteins ⟶ proteoses and peptones
Rennin	Prorennin	,, ,,	,,	Pepsin	Curdles milk
Lipase		Pancreas	Small intestine		Fats ⟶ fatty acids and glycerol
Pancreatic amylase		,,	,, ,,	Enterokinase and trypsin	Starch and dextrin ⟶ maltose
Trypsin	Trypsinogen	,,	,, ,,	Trypsin	Proteins, proteose and peptones ⟶ smaller units including amino acids
Chymotrypsin	Chymotrypsinogen	,,	,, ,,		
Enterokinase		Small intestine	Small intestine	Trypsin	Activates trypsinogen to trypsin
Carboxy-peptidase	Procarboxy-peptidase	,, ,,	,, ,,		Proteoses and peptones ⟶ amino acids
Amino-peptidase	Proamino-peptidase	,, ,,	,, ,,	,,	Dipeptides ⟶ amino acids
Dipeptidase	Prodipeptidase	,, ,,	,, ,,	,,	Maltose ⟶ glucose
Maltase		,, ,,	,, ,,		Sucrose ⟶ glucose and fructose
Sucrase		,, ,,	,, ,,		Lactose ⟶ glucose and galactose
Lactase		,, ,,	,, ,,		

DIET

The rat is **omnivorous**. Wild rats will scavenge almost any sort of food, not only those stored for man and his domestic animals but also such materials as cotton waste and garbage of any kind.

    The dietary requirements of the albino rat have been very carefully studied. A good mixed diet will contain all the essential substances in at least minimal quantities, besides **roughage**, which is indigestible and is voided as the faeces.

    As in the case of other animals, the essential substances may be classified as **organic** and **inorganic**.

ORGANIC SUBSTANCES IN THE DIET

    The organic components of the diet are **carbohydrates**, **fats**, **proteins** and **vitamins**.

    Carbohydrates, fats and proteins are all taken in relatively large quantities and, when oxidized, yield energy for the vital activities. Proteins are also used for the construction of protoplasm for growth and repair. Vitamins are taken in relatively small quantities and, though they do not yield energy, are essential for health and normal development.

(1) CARBOHYDRATES

    The quantities of carbohydrates in the diet can be varied considerably. There is no reason to believe that any form of carbohydrate is essential to the rat, though normally it makes up a very large proportion of the food. On oxidation, carbohydrates yield energy for metabolic activity.

(2) FATS

    On oxidation **fatty acids** yield more than twice as much energy as the equivalent weight of carbohydrate or protein. The rat can tolerate a very large amount of fat. If fats are deficient, certain amounts of saturated fatty acids and neutral fats can be synthesized from other sources, but the higher **unsaturated fatty acids** cannot be synthesized and are therefore **essential** in the diet. Deficiency of the essential fatty acids produces scaly skin, loss of hair, inflamed and swollen tail, disturbances of fluid balance, loss of reproductive capacity and ultimately death.

(3) PROTEINS

    During digestion the long protein chains are split into their constituent **amino acids**, but after absorption they can be built up again into the different proteins needed for growth and repair of the animal's own tissues. Of the 22 amino acids which occur in nature, **valine**, **leucine**, **isoleucine**, **lysine**, **methionine**, **phenylalanine**, **threonine** and **tryptophan** are **essential** because they cannot be synthesized in the mammalian body. **Histidine** is also essential to the rat and **arginine** promotes growth because it is not synthesized in sufficient quantities. The other amino acids are interconvertible provided the total supplies are adequate to provide the necessary amino groups. Excess and unwanted amino acids are **deaminated**, i.e. the nitrogenous portion is removed. The organic portions of the molecules are utilized to liberate energy or are converted into carbohydrate or fat.

Optimum growth and lactation of the rat occur when the diet has between 25% and 30% protein. Higher protein diets (up to 40%) produce improved fertility but no better growth or lactation, while lower protein diets (down to 16%) produce the same gains in body weight, though a higher proportion of the gain is fat.

(4) VITAMINS

Vitamins are organic compounds of varied constitution. Some of them have been synthesized but others are obtainable only from organic sources. A substance which is a vitamin to one animal is not necessarily so to another, the difference depending on whether the animal is dependent on supplies in the diet or whether it can manufacture sufficient in its own tissues.

(i) Vitamin A

Vitamin A is fat-soluble. It is essential for growth, normal vision and maintenance of the mucous membranes of the body. Deficiency causes stunting, "night blindness" and cornification of the respiratory passages, lungs, alimentary canal, salivary glands and urinogenital tract. The incisor teeth become pale and deformed.

Vitamin A is available in certain oily foods, but can also be obtained from **carotene**, the orange pigment of plants. Carotene is not utilized as readily as vitamin A itself and therefore about four times as much of it is required to have the same effect. The potency of vitamin A preparations is measured in "International Units" and is tested by the effect produced on experimental animals, including rats.

(ii) Vitamin B

Vitamin B is not a single substance as was at first thought. It has been shown to consist of a group of unrelated substances which commonly occur together in nature and form the vitamin B complex. Some of these substances have been extracted and purified, while others are known only by their effects.

Vitamin B$_1$ or **thiamin** is essential for carbohydrate metabolism because it actually takes part in the formation of one of the respiratory enzymes. Disturbances of the nervous system known as **polyneuritis (beri-beri)** are secondary effects of deficiency. The amounts of thiamin needed for health are proportional to the non-fat calories in the diet as a whole.

Vitamin B$_2$ or **riboflavin** is also an essential constituent of one of the respiratory enzymes. Deficiency produces symptoms of **dermatitis**, stunts the growth of the young rats and frequently produces a form of cataract.

Nicotinic acid is an essential substance in the diet of man, but does not appear to be needed by rats. It is probable that the rat can synthesize enough for its requirements and it therefore cannot suffer from **pellagra**. Excess nicotinic acid in the diet of rats is detrimental.

Other constituents of the vitamin B complex are B$_{12}$, choline, pyridoxine, pantothenic acid, biotin, inositol, para-aminobenzoic acid and folic acid. **Vitamin B$_{12}$** is the antianaemic factor essential for the maturation of new red blood corpuscles. **Choline** is believed to be concerned with the metabolism of sulphur containing amino acids. Deficiency in the diet of rats produces renal lesions, fatty liver and paralytic symptoms in the nursing young. Though this substance is also essential in the metabolism of man, to him it is not a vitamin because he can synthesize sufficient for his own requirements. Deficiency of **pyridoxine** and **pantothenic acid** similarly have no known effects on man though treatment has been known to help some sufferers from pelagra and beri-beri symptoms. Pyridoxine deficiency causes a form of anaemia in dogs and pigs and pantothenic acid deficiency causes cardiac, adrenal and skin defects in rats. Deficiency of **biotin** causes an interesting form of dermatitis known as "egg white injury" because it is a specific effect of the removal of biotin from the diet by raw white of egg if taken in large quantities. Deficiency of **inositol** arrests the growth of young rats and produces baldness. Medicinally this substance is thought to help certain cases of psoriasis and in the utilization of vitamin E. Deficiency of **para-aminobenzoic acid** causes greying of hair in rats but has no known effect on man. **Folic acid** is essential for the synthesis of desoxyribose nucleic acid, an essential component of all living cells.

(iii) Vitamin C

Vitamin C is the antiscorbutic vitamin which prevents **scurvy** in man. The rat appears to be able to synthesize a sufficient quantity of this substance in its own body and shows no adverse symptoms when deprived of it in its diet. Indeed, only the primates (man and the monkeys) and the guinea-pig have been found to suffer from scurvy. Vitamin C is essential for the making of new red corpuscles.

(iv) Vitamin D

Vitamin D is fat-soluble and is the antirachitic vitamin, i.e. it helps to prevent the bone deformity known as **rickets**. It is unable to do so if calcium or phosphorus are deficient. If the supplies of calcium and inorganic phosphorus are adequate and the Ca:P ratio is between 1:1 and 2:1, the rat does not develop rickets even with the total absence of vitamin D from the diet; but the presence of vitamin D enables it to tolerate a considerable divergence from the optimum Ca:P ratio without developing rickets. If, however, the total supplies of phosphorus are low, the addition of vitamin D may stunt growth. It is believed that this effect is due to the conversion of organic phosphorus to inorganic phosphorus by the vitamin, so that the soft tissues are deprived of an essential constituent for their development.

A substance closely similar to vitamin D and having similar effects can be manufactured in the body under the action of **ultraviolet light**; therefore in all experiments on the potency of vitamin D preparations such light must be excluded.

(v) **Vitamin E** or **alpha-tocopherol**

Deficiency of vitamin E reduces the fertility of rats by causing sterility in the male and failure of formation of the allanto-chorionic placenta at the twelfth day, so that the foetuses die and are resorbed. The effect of deficiency in the young is to reduce growth and produce degeneration of skeletal muscle with partial paralysis. Such effects are unknown in man, though supplementary vitamin E has been given to assist pregnancy in cases where there has been repeated miscarriage.

(vi) **Vitamin K**

Vitamin K deficiency in birds results in a tendency to internal haemorrhage and extension of the time taken for blood to clot. This effect is probably due to a reduction in the amount of prothrombin present in the blood. There is evidence that rats can develop severe symptoms but that these may take time to appear. Vitamin K is fat soluble and dietary deficiency is rare in man except when fats cannot be digested and absorbed, e.g. in conditions of obstructive jaundice. In such cases the same symptoms are seen as in birds.

INORGANIC SUBSTANCES IN THE DIET

The diet includes a number of **inorganic salts** and **water**.

SALTS

A number of different inorganic salts are necessary for correct development and functioning of the tissues. Though these are all absorbed in combined state as mineral salts, it is usual to consider the amounts of each constituent element separately, irrespective of the other parts of the molecules in which it may be supplied.

Calcium and phosphorus are needed for bone formation and also for growth and functioning of the other tissues. The ratio of the amount of calcium to the amount of inorganic phosphorus in the diet is far more important than the actual quantities of either substance.

If a rat is given little phosphorus and vitamin D is also deficient, **rickets** develops very rapidly. If phosphorus alone is deficient, such as is available is used for bone formation, and growth of the soft tissues is inhibited. Should the deficiency be prolonged, the calcium passing through the alimentary canal removes phosphorus from the body. This phosphorus is mobilized from the skeleton and the corresponding amounts of calcium are excreted in the urine.

If calcium alone is deficient, there are no specific symptoms, but growth is retarded because the food materials are less efficiency utilized.

The availability of the calcium and phosphorus in the diet of mammals varies. When oxalates are present, as in spinach, calcium is precipitated and is not absorbed. It is also precipitated by phytic acid which, though it contains phosphorus, is useless as a source of this element. Phytic acid occurs in many grains and other seeds whose phosphorus content is therefore unavailable from the nutritional point of view, but it does not occur in leaves and stems, the phosphorus of which is in inorganic form.

Potassium is essential for life, as it is a normal component of all cells. In the rat, deficiency produces severe ill health and rapid death.

Sodium is also essential for life. In the rat, deficiency produces reduced appetite and other disturbances leading eventually to death. The effects are not as rapid as in the case of potassium, and the minimum for normal growth is much less.

Chlorine is needed for growth, but in the rat, deficiency has less effect than deficiency of sodium. If chlorine and sodium are deficient at the same time, the effect of the sodium deficiency is reduced.

Iron and copper are both needed for the formation of haemoglobin for the red blood corpuscles. The iron forms an actual component of the haemoglobin while the copper catalyses its synthesis. The total amount of iron in the body is small, but a certain amount can be stored in the liver and spleen and utilized as required. The availability of the iron in the diet varies. As in the case of calcium, phytic acid prevents it from being absorbed.

Besides assisting in the manufacture of haemoglobin, copper is incorporated in the molecules of two of the enzymes used in the oxidation of carbohydrates. In the rat, deficiency thus impairs carbohydrate metabolism at the same time as it causes anaemia.

Iodine is an essential component of the **thyroxin** manufactured by the thyroid gland. A very small amount of iodine is needed in the diet, but deficiency results in goiterous swelling of the gland to four or more times its normal weight.

Magnesium, manganese, zinc and **cobalt** have been shown to be essential in the diet of rats, but in very small quantities only, so that they are rarely deficient. Excess of zinc or cobalt is toxic.

Fluorine occurs in bones, teeth and other tissues, but has not been shown to be essential to rats.

WATER

Adequate supplies of water are essential to make good the inevitable losses through **excretion**, **respiration** and **perspiration**.

The kidneys of mammals are designed to conserve water to a certain extent, but the nitrogenous waste is still removed in solution. The faeces are partially dried by absorption of water in the large intestine, but nevertheless they are still moist. The amount of moisture lost through breathing depends on the amount present in the inspired air, because the expired air is always saturated with water vapour at body temperature. The amount of water lost in this way is considerable. Loss of water by perspiration is very important in some mammals; it depends on the number of sweat glands and their activity (see page 49).

A certain amount of water is present in almost all food, but in addition the rat must drink to satisfy its requirements.

VASCULAR SYSTEM

As in all vertebrates, the **blood** and **lymphatic** systems are interrelated. Lymph is formed from tissue fluid which is itself an exudate from the blood-vessels in the tissues. It is returned to the blood-stream at definite points.

STRUCTURE OF BLOOD

Blood is a slightly sticky reddish fluid. That of the rat is of specific gravity 1·056 and pH 6·7–7·0. Whole blood consists of a liquid component called **plasma** and cells called **corpuscles**.

PLASMA

Plasma is yellowish in colour and contains **proteins**, **salts**, **food** and **waste materials** in solution. In the rat the dissolved solids amount to about 8% of the plasma while the other 92% is water.

The **blood proteins** include **serum albumin**, **serum globulin** and **fibrinogen**. They are in colloidal solution and make the plasma viscous. The fibrinogen is converted into **fibrin** during clotting.

The **blood salts** maintain a constant osmotic environment for the tissues and by their buffering action maintain a relatively constant pH. Some of them also have specific effects, e.g. the role of calcium ions in clotting.

Food materials are carried in the plasma to the tissues and **waste materials** are carried away from them to the excretory organs.

The plasma also contains dissolved **gases**, oxygen, nitrogen and carbon dioxide, and minute amounts of **enzymes**, **hormones**, **antibodies** and **antitoxins**—see *Functions of Blood*, page 40.

CORPUSCLES

The blood cells are of three different kinds, red corpuscles or **erythrocytes**, white corpuscles or **leucocytes**, and **blood platelets**.

ERYTHROCYTES

Mammalian red corpuscles are disc-shaped. Each corpuscle consists of an **envelope** containing spongy cytoplasm called **stroma**. When first formed in the red bone marrow it has a nucleus, but this disintegrates when it is liberated into the blood-stream, so that the sides of the disc collapse and it becomes **biconcave**.

The stroma contains the red pigment **haemoglobin**. Haemoglobin is a conjugated protein, i.e. it consists of **globin**, which is protein united with **haematin** which is a complex organic substance containing **iron**. Haemoglobin can combine reversibly with oxygen and with carbon dioxide. **Oxygen** becomes attached to the haematin part of the molecule, forming **oxyhaemoglobin**. The equilibrium represented by the equation *haemoglobin* $+ O_2 =$ *oxyhaemoglobin* is governed by the amount of oxygen present, thus oxyhaemoglobin tends to be formed in the lungs and oxygen to be liberated in the tissues. **Carbon dioxide** becomes attached to the amino groups of the protein portion of the haemoglobin to form **carbaminohaemoglobin (carbhaemoglobin)**. This reaction is greatly influenced by the degree of oxygenation of the haemoglobin as well as the amount of carbon dioxide present, so that the formation of oxyhaemoglobin in the lungs helps to liberate carbon dioxide from the carbamino state. The greater part of the carbon dioxide carried by the blood is however buffered as **bicarbonate**.

Thus the red corpuscles with their haemoglobin are responsible for the transport of oxygen from the lungs to the tissues and of much of the carbon dioxide from the tissues to the lungs.

Note. Haemoglobin can combine with **carbon monoxide** to form **carboxyhaemoglobin**. Though the reaction is reversible the equilibrium constant is such that carboxyhaemoglobin is formed even at very low concentrations of carbon monoxide, thus blocking the use of the haemoglobin as a carrier of oxygen and producing a form of asphyxiation of the tissues known as carbon monoxide poisoning.

After a few weeks in the circulation the red corpuscles become worn out and are destroyed. Destruction takes place chiefly in the spleen. Iron from the haemoglobin is passed to the liver and stored there until it is required for the manufacture of new red corpuscles in the red bone marrow.

The average **number** of erythrocytes in rat blood is 9–9½ million per mm³, with rather more in the female than the male. (Human: 4½–5 million per mm³, with more in ♂ than ♀.)

The average **size** of the erythrocytes of the adult rat is 6·3 μm in diameter and of the foetus 10 μm in diameter. (Human: 7·0 μm in diameter.)

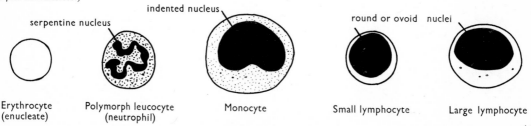

Fig. 57 **Blood corpuscles of rat**

c. × 1600

LEUCOCYTES

White corpuscles are nucleated and lack haemoglobin. They are of variable size and shape and are classified according to the form of the nucleus and the staining reactions of the cytoplasm and its granules. The types and the average percentages of each type in the rat and in man are listed below.

Rat		*Man*	
Polymorph granulocytes (neutrophil)	20%	Polymorph granulocytes (neutrophil)	66%
Lymphocytes	76%	Lymphocytes	25%
Monocytes	1%	Monocytes	5%
Eosinophil	2%	Eosinophil	3%
Basophil	1%	Basophil	1%

The total number of white corpuscles in the rat is 6,000–18,000 per mm³ of blood. (Human, 5,000–10,000 per mm³.) **Lymphocytes** are formed in the **lymph nodes**, spleen and thymus—see page 61—while the other **leucocytes** are formed in the **red bone marrow**. There is a general increase of all types during infection to combat germs and their poisons. All types except the lymphocytes are capable of amoeboid movement by which they engulf germs and damaged tissue cells. This process is known as **phagocytosis**. They can also crawl along the walls of blood-vessels and squeeze between the cells forming the blood capillaries to pass into the tissues. Lymphocytes are incapable of phagocytosis and are concerned with **immunological response** to invasion by foreign substances collectively called **antigens**. The **antibodies** produced are highly specific and are classified as **lysins**, which dissolve invading organisms; **opsonins**, which promote phagocytosis; **agglutinins**, which cause bacteria and other cells to stick together in clumps; and **precipitins**, which act against soluble antigens causing precipitation. Induced immunity may last for a considerable time after the original invasion.

BLOOD PLATELETS

The blood platelets are enucleate irregularly shaped discs formed by the breakdown of large cells called **megakaryocytes** found chiefly in the red bone marrow. The blood platelets tend to adhere to any wettable or roughened surface, where they form clumps and disintegrate, liberating a substance known as the **platelet factor (thromboplastinogenase)**. This substance plays an important part in the process of clotting.

The average number of platelets in the rat is 800,000 per mm³. (Human, 300,000 per mm³.) The average size of the platelets of the rat is 0·5 μm. (Human, 2–3 μm.)

CLOTTING OF BLOOD

Clotting is a function of the **plasma** and takes place even when the corpuscles have been removed, e.g. by centrifuging.

Clotting occurs when the soluble protein, **fibrinogen**, is converted into the stringy elastic **fibrin** fibres which stick together to form a meshwork. This gradually shrinks, expelling the fluid component of the plasma which is then known as **serum**. A clot formed from whole blood is stiff because corpuscles become entangled in the shrinking fibrin network, but a clot formed from pure plasma is soft and spongy.

Clotting normally occurs only when blood is **shed**. The stimulus which starts the clotting process is a **roughened surface**, e.g. damaged tissue, or a **wettable surface**, e.g. glass. How such a surface acts is still a matter of much discussion, but the effect is to start a chain of reactions which result in the production of **thrombin** which in turn stimulates the conversion of fibrinogen to fibrin.

Thrombin is formed from **prothrombin** by **thromboplastin** (**thrombokinase**, tissue extract) in the presence of **calcium ions** (**Ca^{++}**). Very little prothrombin is free in normal blood, but it is formed from its precursor, prothrombinogen, by the stimulus of a roughened or wettable surface or by thrombin itself, so that, once started, the reaction is **autocatalytic**. Thromboplastin is not free in blood plasma but is present in **cells** from which it is liberated when tissue is damaged. It is also formed as a result of disintegration of blood **platelets** when they clump and adhere to a roughened or wettable surface or when they are acted on by thrombin.

In its simplest form the clotting process may be expressed thus:

Preparatory reactions
- prothrombinogen+roughened or wettable surface or thrombin ⟶ **prothrombin**
- platelets+roughened or wettable surface or thrombin ⟶ clumping and disintegration
- damaged tissue or disintegration products of platelets ⟶ **thromboplastin** (thrombokinase)

Key reaction   PROTHROMBIN+THROMBOPLASTIN+Ca^{++} ⟶ THROMBIN

Effective reaction   FIBRINOGEN+THROMBIN ⟶ FIBRIN ⟶ CLOT

Corpuscles, if present, become entangled in the fibrin and the clot gradually shrinks, expelling serum.

The time taken for rat blood to clot is 2½ minutes, while for human blood it is 7 minutes. This difference is due to the greater concentration of fibrinogen in rat than in human blood.

Clotting does not normally take place while the blood is in circulation in the body because **heparin** formed by the liver and other tissues, including the basophil leucocytes and walls of the blood vessels, prevents the platelets from sticking to these walls and acts as an **antiprothrombin** and an **antithrombin**, inhibiting the production of thrombin from prothrombin and the effect of any thrombin present.

Note. Various **accelerator factors** have been postulated by different workers, e.g. the **serum accelerator** (serum Ac-globulin) which is said to be made active by small amounts of thrombin and then assist the production of more thrombin, and at least two **platelet factors** with different actions. Until the matter is more clearly understood it is advisable for the majority of students to ignore these and to concentrate on the essential reactions as described above. It is worth noting, however, that it may eventually be proved that the whole process is a series of physical changes in a colloidal system rather than a series of chemical reactions as is generally supposed.

BLOOD GROUPS

The red corpuscles may carry substances known as **agglutinogens** which cause them to clump together when they encounter the corresponding **agglutinins** in the blood serum. In both the rat and man there are four major blood groups, those with agglutinogen **A**, those with agglutinogen **B**, those with agglutinogens **A and B** and those with **neither**. They are known as *group A, group B, group AB* and *group O* respectively.

Corpuscles carrying agglutinogen **A** are clumped by agglutinin α, but not by agglutinin β, and vice versa. In the rat and in man agglutinin α occurs in the serum of *group B* and *group O*, but not in *group A* and *group AB*. In the rat agglutinin β does not occur naturally but can be produced in response to injection of blood corpuscles carrying agglutinogen **B** into *group A* or *group O* animals. Agglutinin β occurs naturally in man.

The blood groups are governed by genetic factors and are therefore hereditary. They are of medical importance during blood transfusion. The corpuscles coming from the donor must not be clumped by the serum of the recipient. The serum of the donor has a negligible effect on the corpuscles of the recipient. As shown in the table below, *group AB* is the universal recipient, while *group O* is the universal donor.

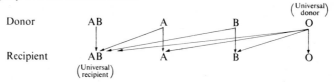

QUANTITY OF BLOOD

The total amount of blood in the rat is very small. The average is 4·3 ml per 100 g of body weight. A loss of even 1 ml may produce **anaemia**. The human body, on the other hand, has about 8·3 ml per 100 g of body weight. When, however, these figures are compared with the average erythrocyte count it is found that there are approximately the same number of red corpuscles per 100 g in the two different species, i.e. the oxygen-carrying power is almost identical.

$$\text{Rat:} \quad \text{red cells per 100 g} \quad 4\cdot3 \times 1,000 \times 9\cdot35 = 40,205 \text{ million}$$
$$\text{Human: red cells per 100 g} \quad 8\cdot3 \times 1,000 \times 5 \quad = 41,500 \text{ million}$$

FUNCTIONS OF BLOOD

Blood is the main **transport system** of the body and also helps to **protect** the body against **disease**.

TRANSPORT

(i) The blood carries **food materials** from the alimentary canal to the tissues where they can be utilized or stored.

(ii) The blood carries **oxygen** from the lungs to the other parts of the body. Though there is some oxygen in solution in the blood plasma the greater part of the oxygen carried is combined with the haemoglobin in the red corpuscles. Where the partial pressure of oxygen is high, i.e. in the lungs, haemoglobin becomes oxyhaemoglobin; but where the partial pressure of oxygen is low, i.e. in the tissues (since oxygen is used up for tissue respiration), the **oxyhaemoglobin** breaks down and the haemoglobin is ready to pick up a fresh supply of oxygen when it is returned to the lungs.

(iii) The blood carries **waste materials** from the tissues to the places where they are excreted. It carries **urea** to the kidneys and **carbon dioxide** to the lungs. The latter is carried chiefly as **bicarbonate** and as **carbaminohaemoglobin**.

(iv) The blood transports **hormones** from the endocrine organs to the tissues which are affected by these chemical messengers.

(v) The blood distributes **heat** from the more active organs to the less active ones and thus helps to maintain an even body temperature in the different parts.

PROTECTION

(i) The blood helps to maintain the tissues in healthy condition by supplying them with their material needs and with the conditions best suited to their life. In this respect the constancy of the **osmotic pressure** and the **pH** of the blood are very important.

(ii) The white corpuscles actively fight invading organisms. The lymphocytes produce **antibodies** and **antitoxins** which act against germs and their poisons while the **phagocytes** actively engulf them and destroy them.

(iii) The **clotting** of blood closes wounds and thus arrests haemorrhage as well as sealing the cut against the entry of germs.

CIRCULATION OF THE BLOOD

Blood flows in a closed system of vessels, the **arteries**, **veins** and **capillaries**. It is kept moving by the action of the **heart**. The right and left sides of the heart are completely separated from one another, so that the circulation to the body (**systemic circulation**) and that to the lungs (**pulmonary circulation**) are entirely distinct.

ARTERIES

Arteries carry blood **away from** the heart. The systemic arteries carry **oxygenated** blood with **oxyhaemoglobin** in the red corpuscles while the pulmonary arteries carry **deoxygenated** blood. The distribution of the main systemic arteries of the rat is shown in Fig. 59. The finer branches are unimportant to most students, but when dissecting it is of interest to notice the very large number of anastomoses, especially in the vessels supplying the viscera. This arrangement is a safety measure found in all vertebrates and allows the parts to be supplied in a roundabout way should there be blockage of the normal channels. The finest branches of the arteries are known as **arterioles** and open into the capillaries.

The walls of the arteries contain **smooth muscle** fibres and **elastic tissue** which maintain pressure on the blood. The pressure in the main arteries of the rat is between 90 and 101 mm. of mercury and, as in man, tends to increase with age. The pressure is lower in the branch vessels and in the arterioles. The walls of the latter contain relatively more muscle and less elastic tissue. This muscle is controlled by **sympathetic nerve fibres** and **hormones** of two types—**vasoconstrictor** and **vasodilator**. The vasoconstrictors produce a state of tone in the walls of the arterioles which restricts the passage of blood. The vasodilators inhibit the vasoconstrictors and thus allow more blood to flow. These neural and hormonal effects may be localized, when they bring about redeployment of blood, or general, when they affect overall blood pressure by altering resistance to flow of blood from the arteries. The lowering of blood pressure as a result of vasodilation is normally very temporary and rapidly adjusted by change in strength of heart beat. Vasoconstriction plus increase in cardiac output result in rise in blood pressure which may be maintained throughout a period of stress, e.g. as a response to fear or anger ("fight or flight")—see page 61.

Rats can suffer from **arteriosclerosis** (hardening of the walls of the arteries) accompanied by abnormally high blood pressure.

VEINS

Veins carry blood **towards** the heart. The systemic veins carry **deoxygenated** blood but the pulmonary veins carry **oxygenated** blood. The distribution of the main systemic veins is shown in Fig. 60. Like the arteries, the veins anastomose. In many places the arteries and veins accompany one another, the chief exceptions being the vessels inside the skull and the **portal system** from the alimentary canal, pancreas and spleen to the liver. The finest tributaries of the veins are known as venules. They receive blood from the capillaries.

The walls of the veins are much thinner than those of the corresponding arteries. They have much less muscle and elastic tissue. The pressure in the main veins of the rat is -90 to -160 mm of water ($\equiv -0.66$ to -1.18 mm of mercury). Veins have **valves** at intervals along their length to help to maintain the flow of blood in the correct direction.

CAPILLARIES

Capillaries form networks in the tissues. They have very thin walls, through which there is easy diffusion of materials from the blood to the **tissue fluid** and thence to the tissue cells and vice versa. Tissue fluid itself is formed from the blood in the capillaries, but at a very slow rate. Abnormal permeability of the blood capillaries without corresponding rapid removal of the tissue fluid as lymph produces the condition known as *oedema*. The average capillary blood-pressure in the rat is 12–21 mm of mercury.

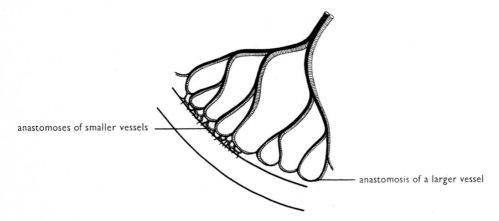

anastomoses of smaller vessels

anastomosis of a larger vessel

Fig. 58 **Anastomosing blood-vessels in the mesentery of the intestine**

Notice that the external
carotid arteries are medial
to the internal carotid
arteries

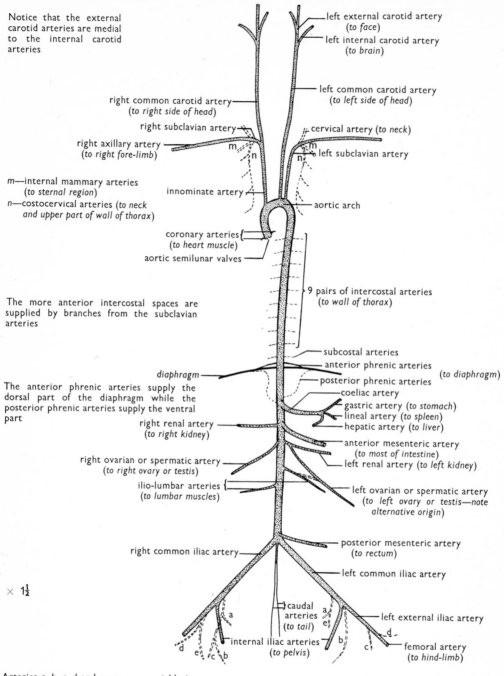

left external carotid artery
(to face)

left internal carotid artery
(to brain)

left common carotid artery
(to left side of head)

right common carotid artery
(to right side of head)

right subclavian artery

cervical artery (to neck)

right axillary artery
(to right fore-limb)

left subclavian artery

m—internal mammary arteries
(to sternal region)

n—costocervical arteries (to neck
and upper part of wall of thorax)

innominate artery

aortic arch

coronary arteries
(to heart muscle)

aortic semilunar valves

9 pairs of intercostal arteries
(to wall of thorax)

The more anterior intercostal spaces are
supplied by branches from the subclavian
arteries

subcostal arteries

anterior phrenic arteries

diaphragm

posterior phrenic arteries

(to diaphragm)

coeliac artery

The anterior phrenic arteries supply the
dorsal part of the diaphragm while the
posterior phrenic arteries supply the ventral
part

gastric artery (to stomach)

lineal artery (to spleen)

hepatic artery (to liver)

right renal artery
(to right kidney)

anterior mesenteric artery
(to most of intestine)

left renal artery (to left kidney)

right ovarian or spermatic artery
(to right ovary or testis)

left ovarian or spermatic artery
(to left ovary or testis—note
alternative origin)

ilio-lumbar arteries
(to lumbar muscles)

posterior mesenteric artery
(to rectum)

right common iliac artery

left common iliac artery

× 1½

caudal
arteries
(to tail)

left external iliac artery

internal iliac arteries
(to pelvis)

femoral artery
(to hind-limb)

Arteries a, b, c, d and e are very variable in
position, often differing on the two sides of
the same rat
    a—vesical arteries (to bladder)
    b—haemorrhoidal arteries (to anal canal)
    c—pudendal arteries (to pubic region)
    d—epigastric arteries (to abdominal wall)
    e—uterine arteries (to uterus in female only
        —anastomose with the ovarian arteries)

Fig. 59 **Principal arteries of the rat**

===== less important vessels which are nevertheless often
observed during dissection

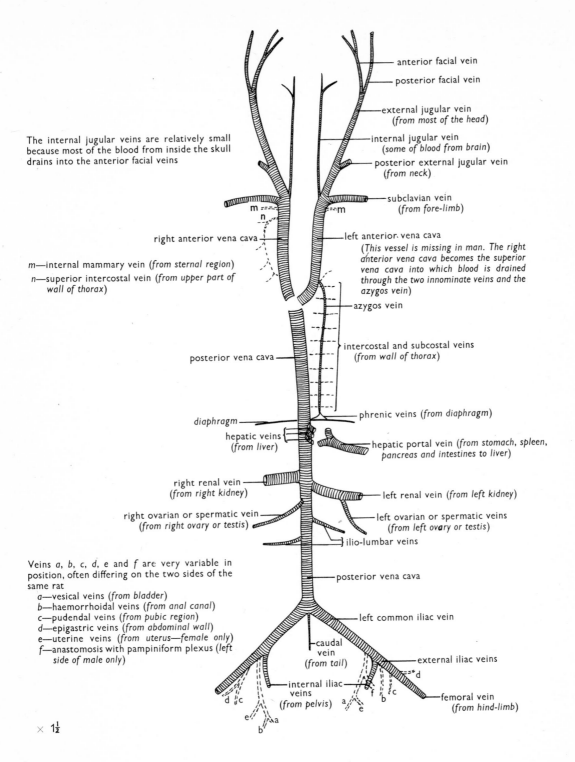

anterior facial vein

posterior facial vein

external jugular vein
(*from most of the head*)

internal jugular vein
(*some of blood from brain*)

posterior external jugular vein
(*from neck*)

The internal jugular veins are relatively small because most of the blood from inside the skull drains into the anterior facial veins

subclavian vein
(*from fore-limb*)

m

n

right anterior vena cava

left anterior vena cava
(*This vessel is missing in man. The right anterior vena cava becomes the superior vena cava into which blood is drained through the two innominate veins and the azygos vein*)

m—internal mammary vein (*from sternal region*)

n—superior intercostal vein (*from upper part of wall of thorax*)

azygos vein

intercostal and subcostal veins
(*from wall of thorax*)

posterior vena cava

phrenic veins (*from diaphragm*)

diaphragm

hepatic veins
(*from liver*)

hepatic portal vein (*from stomach, spleen, pancreas and intestines to liver*)

right renal vein
(*from right kidney*)

left renal vein (*from left kidney*)

right ovarian or spermatic vein
(*from right ovary or testis*)

left ovarian or spermatic veins
(*from left ovary or testis*)

ilio-lumbar veins

Veins *a, b, c, d, e* and *f* are very variable in position, often differing on the two sides of the same rat
  a—vesical veins (*from bladder*)
  b—haemorrhoidal veins (*from anal canal*)
  c—pudendal veins (*from pubic region*)
  d—epigastric veins (*from abdominal wall*)
  e—uterine veins (*from uterus—female only*)
  f—anastomosis with pampiniform plexus (*left side of male only*)

posterior vena cava

left common iliac vein

caudal vein
(*from tail*)

external iliac veins

d

internal iliac veins
(*from pelvis*)

f c

a
e

b

femoral vein
(*from hind-limb*)

d c

e a

b

$\times\ 1\frac{1}{2}$

===== less important veins which are nevertheless often observed during dissection

Fig. 60 **Principal veins of the rat**

HEART

The heart lies in the ventral part of the thorax between the lungs. It is enclosed in a double-walled membranous bag, the **pericardium** (see page 25).

The heart has four chambers, two **atria** and two **ventricles**. (*Note.* The atria are frequently called auricles, but this term is used for small parts only of the atria in man.) The right atrium communicates with the right ventricle and the left atrium with the left ventricle, but the two sides of the heart are separated by a complete thick **septum**. The septum is curved so that the right ventricle is wrapped round part of the left ventricle.

Between each atrium and the corresponding ventricle is an **atrio-ventricular valve**, the **tricuspid valve** on the right and the **bicuspid** or **mitral valve** on the left. The edges of both valves are attached to tendinous strands, the **chordae tendineae**, which arise from small **papillary muscles** on the walls of the ventricles and prevent the valves from becoming inverted.

The **right atrium** receives **deoxygenated blood** from the body through the three **venae cavae**, openings of which lie on its dorsal wall.

The **right ventricle** forces **deoxygenated blood** into the **pulmonary artery** and thence to the lungs for oxygenation. The artery leaves the antero-ventral region of the ventricle and has **semilunar valves** at its mouth.

The **left atrium** receives **oxygenated blood** from the lungs through the **pulmonary veins** which, in the rat, join before reaching the heart, so that there is only one opening in the dorsal wall of the atrium.

The **left ventricle** forces **oxygenated blood** into the **aorta** and thence round the body. The artery leaves the antero-dorsal region of the ventricle and thus lies dorsal to the pulmonary artery which curves across it. It has three strong **semilunar valves** at its mouth.

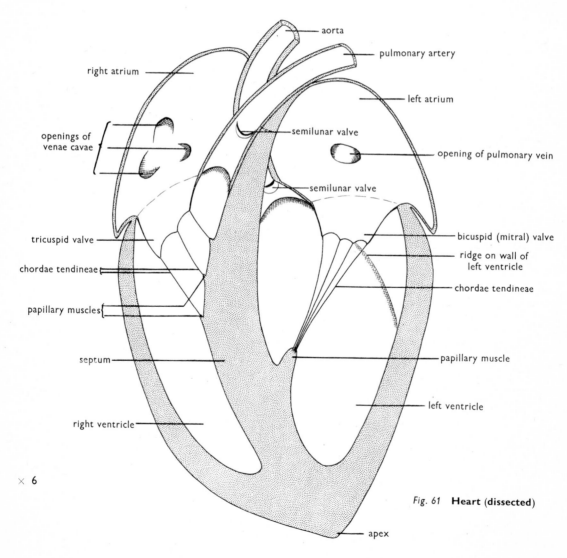

× 6

Fig. 61 **Heart (dissected)**

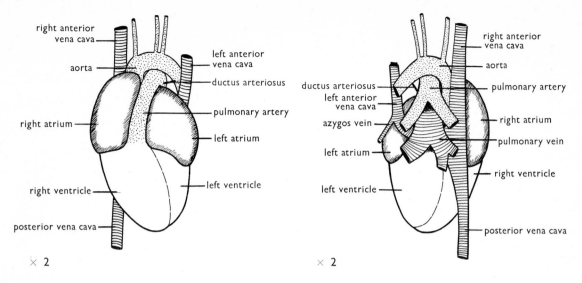

Fig. 62 **Heart (ventral view)**

Fig. 63 **Heart (dorsal view)**

Note. Before birth the septum is incomplete and some of the blood short-circuits from the right atrium to the left. There is also a connection, the **ductus arteriosus**, between the pulmonary artery and the aorta, so that very little blood is sent to the developing lungs, which are as yet not functioning.

At birth, when the lungs fill with air and start to function, the septum becomes complete and the ductus arteriosus closes so that the double circulation is established. The ductus arteriosus becomes a fibrous band which is retained throughout life.

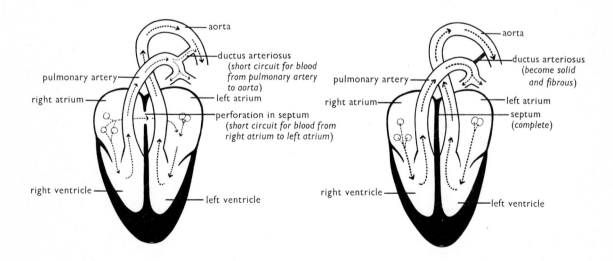

Fig. 64 **Course of circulation through the heart of the foetus before birth**

Fig. 65 **Course of circulation through the heart after birth**

METHOD OF FUNCTIONING OF THE HEART

The heart beats regularly throughout life and pumps blood through the blood-vessels. Each complete beat or **cardiac cycle** consists of a period of **systole (contraction)** and a period of **diastole (relaxation** and **rest)**.

Contraction starts in the dorsal wall of the right atrium and spreads across both atria, closing the apertures of the veins and forcing blood into the ventricles under pressure. The tricuspid and bicuspid valves close as the ventricles start to contract, so that blood cannot flow back into the atria and is forced into the arteries under considerable pressure. As the ventricles contract the papillary muscles contract a compensating amount and thus maintain the tension on the chordae tendineae so that the valves do not turn inside out.

The atria **relax** and **refill** with blood while the ventricles are contracting, so that blood is ready to flow through into the latter when they in turn relax. Thus the heart is never completely empty. When the ventricles relax, backflow from the arteries is prevented by the semilunar valves.

The heart of the rat beats approximately 300 times a minute, i.e. over four times as fast as the average rate of beat in man.

CONTROL OF RATE OF HEART-BEAT

The heart is composed of a special type of muscle known as **cardiac muscle** which has an inherent rhythmic contractility, but the rate of this rhythm is affected by impulses from the **parasympathetic** fibres of the cardiac branches of the **vagus nerves** and the **sympathetic** fibres of the **cardiac nerves** (see pages 67, 68). The parasympathetic impulses inhibit the contraction and slow the heart down, while the sympathetic impulses stimulate it and make the beat more rapid.

Within the cardiac muscle contraction passes in a wave-like manner from fibre to fibre. There is a narrow connective tissue partition between the walls of the atria and those of the ventricles, which prevents direct transmission of the wave of contraction from the atria to the ventricles. A strand of specially sensitive conducting fibres in the septum relays the movement started in the atria to the apex of the heart. Thus contraction of the ventricles follows slightly after that of the atria and forces blood towards the mouths of the great arteries rather than away from them.

LYMPH

Lymph is similar in composition to blood, except that it has no red corpuscles. Its rate of flow is affected by the rate of formation of **tissue fluid**. Like blood it is capable of clotting.

The **lymph vessels** are thin-walled like veins but have even more **valves**. They are interrupted at intervals by lymph nodes.

The **lymph nodes** of the rat are relatively large and very conspicuous during dissection. They lie in definite positions compared with surrounding structures. Their identification is unimportant to the average student, but those of the neck must not be confused with the salivary glands.

The largest lymph vessel, the **thoracic duct,** opens into the junction of the left subclavian and the left internal jugular veins. It is fine and colourless and usually passes unnoticed during dissection. The very short **right lymphatic duct** opens into the junction of the right subclavian and right internal jugular veins.

SPLEEN

The spleen is a bright red body lying close to the greater curvature of the stomach. It contains patches of lymphoidal tissue amongst soft **splenic pulp** in which there are numerous blood-vessels and blood-spaces. It acts as a reservoir for red blood corpuscles and destroys those which are worn out.

Removal of the spleen of the rat produces a drop in the red cell count to 1 million per cub. mm. and the rat dies of anaemia. In man the effect of removal of the spleen is much less serious, and perfect health can be maintained afterwards.

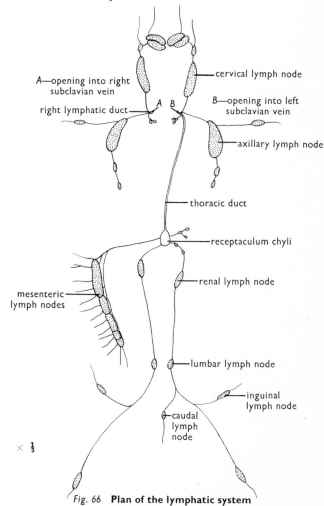

A—opening into right subclavian vein

right lymphatic duct

B—opening into left subclavian vein

cervical lymph node

axillary lymph node

thoracic duct

receptaculum chyli

renal lymph node

mesenteric lymph nodes

lumbar lymph node

inguinal lymph node

caudal lymph node

$\times \frac{1}{3}$

Fig. 66 **Plan of the lymphatic system**

RESPIRATORY SYSTEM

THE respiratory system consists of two **lungs** and the passages by which their internal cavities are connected to the exterior. These **respiratory passages** are the **nasal cavities**, the **pharynx**, the **larynx**, the **trachea** and the **bronchi**.

NASAL CAVITIES

The nasal cavities are separated from one another by the **nasal septum** and from the buccal cavity by the **palate**—see Figs. 11 and 53. Their external openings are the **nostrils**. In the rat these lie on a narrow region of naked skin above the cloven upper lip known as the **rhinarium**. In healthy rats this region is moist and cool. Internally, the nasal cavities open into the **naso-pharynx**.

The **mucous membrane** lining the nasal cavities is **moist** and **ciliated** and contains numerous **mucus glands**. Its area is vastly increased by the convolutions of the turbinate and ethmoid bones—see Fig. 11. Air breathed through the nose is **warmed** to body temperature and **moistened** to saturation by contact with this membrane. Dust from the air adheres to the mucus, which is moved slowly into the pharynx by the cilia and is then swallowed. Thus the air is **cleansed**.

The **olfactory nerves** have numerous naked endings in the mucous membrane of the upper regions of the nasal cavities. By means of these, **odours** of food, etc., can be recognized. The sense of **smell** is very well developed in the rat—see also page 70.

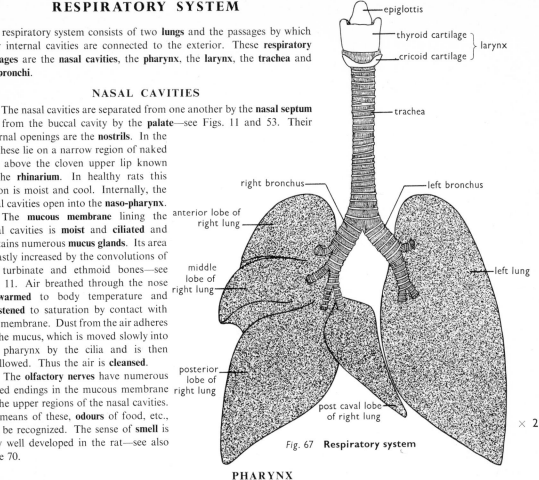

Fig. 67 **Respiratory system**

PHARYNX

The pharynx is divided into the **naso-pharynx** above the palate and the **oro-pharynx** behind the buccal cavity. The latter region is shared by the respiratory and the digestive systems so that air can pass in and out through the mouth as well as through the nose—see Fig. 53. The advantages of nose-breathing lie in the effects of the mucous membrane of the nasal cavities on the condition of the air, as mentioned above.

The edge of the **soft palate** acts as a **valve** to prevent food from passing into the naso-pharynx and thence into the nasal cavities during swallowing. The opening from the pharynx into the larynx is called the **glottis**. During swallowing the glottis is closed and covered by the lid-like **epiglottis**.

LARYNX

The larynx or voice box is the first part of the windpipe. It is supported by a large shield-shaped **thyroid cartilage** and a signet-ring-shaped **cricoid cartilage**.

Inside the larynx lie the **vocal folds**. When these membranes are relaxed air passes out from the lungs over them soundlessly. When they are tensed by contraction of the laryngeal muscles the air movements cause them to vibrate and produce **sounds**, the **pitch** of which varies with the tension. These sounds are magnified by resonance in the buccal and nasal cavities and man can transform them into articulate speech by altering the position of the tongue and the shape of the lips. The rat can utter a variety of squeaks.

TRACHEA AND BRONCHI

The trachea forms the main part of the windpipe. It lies ventral to the oesophagus and is supported by **C-shaped bands** of **cartilage** which keep it open against the pressure from the surrounding organs but allow for stretching as the bolus passes down the oesophagus during swallowing.

The trachea extends into the thorax dorsal to the heart and divides into two **bronchi**, one to each lung. These are supported by complete **rings** of **cartilage** and branch to form the **bronchial tubes** within the lungs. The finest branches are called **bronchioles**.

The trachea, the bronchi, the bronchial tubes and the bronchioles are all lined throughout with **ciliated mucous membrane** in which there are numerous **goblet cells**. The mucus produced by these cells is distributed and moved slowly away from the lungs by cilia. This helps to remove any dust particles which may enter the respiratory passages and to prevent them from accumulating in the lungs.

LUNGS

The lungs lie in the **thorax** on either side of the heart. Each lung lies in a **pleural cavity** which is lined with serous membrane—see page 25. This membrane forms the **parietal pleura** lining the thoracic wall and the **visceral pleura** covering the lung surface.

In the rat the right lung is divided into four lobes, but the left lung is undivided.

The lungs of mammals are bright pink spongy bags whose internal cavities are greatly subdivided to produce numerous **air sacs**, on the walls of which are microscopic pockets called **alveoli**. The airs sacs communicate with the bronchioles and thence, via the main respiratory passages, with the outside air.

The air sacs are bound together by loose connective tissue in which there are numerous blood-vessels. The **capillary blood-vessels** lie very close to the walls of 'the alveoli to facilitate gaseous exchange between the air and the blood.

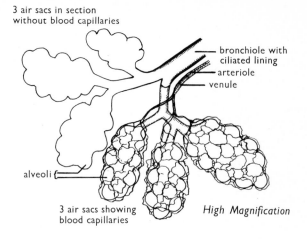

3 air sacs in section without blood capillaries

bronchiole with ciliated lining
arteriole
venule

alveoli

3 air sacs showing blood capillaries

High Magnification

Fig. 68 **Lung structure (highly diagrammatic)**

RESPIRATION

Respiration is the means whereby **food substances** are **oxidized** with liberation of **energy** for body activities.

(a) INTERNAL RESPIRATION

Internal or **tissue** respiration takes place in every living **cell** and involves the oxidation of a variety of organic substances, chiefly **carbohydrates** and **fats**. Any **proteins** used are first deaminated and then their non-nitrogenous portions are treated like carbohydrates and fats. Whatever the material oxidized, **carbon dioxide** and **water** are produced as waste materials, but the proportion of carbon dioxide produced to oxygen utilized varies with the substrate (see *Metabolic rate*, page 50).

(b) EXTERNAL RESPIRATION

External respiration is the means by which oxygen is obtained and carbon dioxide is removed. **Gaseous exchange** takes place by **diffusion** through the very delicate linings of the alveoli of the lungs.

Oxygen from the air in the alveoli dissolves in the thin film of moisture covering the cells, diffuses through the cells into the blood and combines with the **haemoglobin** of the red corpuscles to form **oxyhaemoglobin**. This is then transported to the tissues, where it dissociates to liberate oxygen for tissue respiration and free the haemoglobin to be utilized again.

Carbon dioxide from the tissues is carried back to the lungs. About 5% of it is in simple solution as **carbonic acid**, 2–10% is combined with haemoglobin as **carbaminohaemoglobin (carbhaemoglobin)** in the red corpuscles (see page 38), and the rest is buffered by the blood salts in the form of **bicarbonate**. In the lungs the oxygenation of the haemoglobin helps the breakdown of carbaminohaemoglobin and of the bicarbonates. The latter effect is due to the more strongly acid nature of oxyhaemoglobin than haemoglobin. The carbonic acid thus liberated dissociates readily and carbon dioxide diffuses out into the air.

BREATHING

Breathing is the means by which air in the lungs is exchanged with fresh air from outside. The amount of air exchanged with each breath is relatively small. Some air is always retained in the lungs and the incoming fresh air is mixed with it. The outgoing air is at body temperature and saturated with water vapour, as well as containing carbon dioxide in exchange for some of its oxygen.

Breathing in or **inspiration** is brought about by contraction of the muscles of the **diaphragm** and of the **thoracic wall**. The diaphragm is dome-shaped. Its muscles originate from the wall of the trunk and are inserted on a **central tendon**. As they contract the dome is flattened and thus the thoracic cavity is elongated; at the same time the sternum is swung forwards and the ribs outwards by the contraction of the **intercostal muscles**, so that the diameter of the thoracic cavity is increased. The increase in the capacity of the thorax produces suction which expands the air spaces in the lungs and draws air in through the respiratory passages.

Breathing out or **expiration** is mainly due to **elastic recoil** when the inspiratory muscles relax, but is assisted by the **abdominal muscles** which press the abdominal viscera against the posterior surface of the diaphragm, thus helping to restore its concavity and force air out of the lungs.

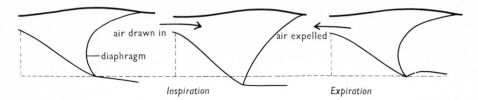

Fig. 69 **Diagrammatic representation of the changes in capacity of the thoracic cavity during breathing**

Breathing movements are controlled through the **respiratory centre** in the **medulla oblongata** of the brain. This centre is made up of the **inspiratory, expiratory** and **inhibitory (pneumotaxic) centres.** Rhythmic breathing is maintained by the **Hering–Breuer reflex.** Motor impulses from the inspiratory centre integrate the action of the inspiratory muscles. At the height of inspiration, tension stimulates sense endings in the walls of bronchioles, air sacs and alveoli. Impulses pass through the vagus nerves to the expiratory centre which inhibits further inspiratory action and causes expiration. Stimuli affecting similarly distributed deflation receptors are then relayed to the inspiratory centre which brings about the next inspiration. The inspiratory centre is also connected to the pneumotaxic centre which relays impulses to the expiratory centre. This in turn inhibits the inspiratory centre if activity of the latter is over-prolonged.

Rate of breathing is adapted to changing need. **Chemoreceptors** in the **carotid bodies** (close to the carotid sinuses) and the **aortic body** (on the inner curve of the aortic arch) sense decrease in oxygen, and increase in carbon dioxide and acidity. The undersurface of the medulla is also directly sensitive to increase in carbon dioxide. Impulses from these receptors result in increased breathing rate, e.g. during and immediately after exercise and at high altitudes.

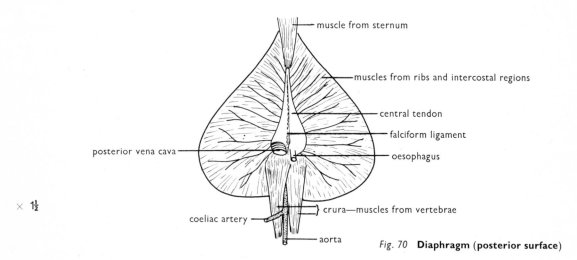

$\times\ 1\frac{1}{2}$

Fig. 70 **Diaphragm (posterior surface)**

METABOLISM

Metabolism is the total of all the chemical activities which go on inside the body. Those processes which are constructive are said to be **anabolic**, while those which are destructive are **katabolic.** All these activities are controlled by **enzymes** which are highly specific and work best in a very limited range of conditions. Maintenance of these conditions in the internal environment is called **homeostasis.**

METABOLIC RATE

Metabolic rate is measured in terms of **heat energy equivalent.** Energy is made available by the oxidation of organic materials —see page 24. Complete oxidation of carbohydrate yields 18 joules per gram, fat 39 joules and protein an average of 18 joules per gram.

The most convenient method of measuring the metabolic rate is by recording the total **oxygen uptake** in a given period and the **respiratory quotient (R.Q.),** i.e. the ratio of the amount of carbon dioxide produced to the amount of oxygen utilized. The respiratory quotient varies with the type of food, being 1·0 for carbohydrate, 0·703 for fat and 0·80–0·82 for protein.

The average oxygen uptake of the rat is 185 mg per 100 g body weight per hour, which allowing for the varying respiratory quotient represents a very high overall metabolic rate of about 84 joules per 100 grams per 24 hours. This rate varies with temperature, season, sex and age, being as much as 70% greater in young than in adult rats. It is also greater morning and evening than in the middle of the day and night.

METABOLISM AND DIET

When the animal is not fasting, the R.Q. of the rat varies from 0·754 to 1·07, with an average of 0·894, i.e. slightly higher than the corresponding average for man (0·85). The very high R.Q. quoted occurs when the diet is almost exclusively carbohydrate and is correlated with a very high total metabolic rate. This is believed to be due to the conversion of some carbohydrate into fat.

On a normal diet including some fat, rats require very little protein. It has been shown that young rats can maintain development when protein is as low as 3% of the food mixture. As in all mammals, the quality of the protein is extremely important. The provision of certain amino acids is essential, though others can be synthesized.

Conservation of protein in the rat is possibly due to the possession of **multilocular adipose** tissue. This tissue forms a pinkish fatty mass between the shoulder-blades and extending round the neck. Its alternative name is the hibernating gland, but this is a misnomer, because the tissue occurs in several mammals such as the rat which do not hibernate, and does not occur in all of those which do. Operations for its removal are almost always rapidly fatal, but this may be due to the extensive nature of the operation rather than to any essential function which it may perform.

When rats are starved for forty-eight hours, all the carbohydrate is used up and the R.Q. falls to an average of 0·725. In conjunction with estimates of nitrogenous excretion, this shows that 90% of the oxidation is of fat and 10% is of protein. In man too much fat in the diet produces ketosis, with ketone bodies, i.e. substances like acetone, in the blood because the naturally occurring fatty acids cannot be fully oxidized unless carbohydrate is being oxidized at the same time. The rat appears to be able to stand a much greater proportion of fat in the diet, but it is noteworthy that more than the normal amount of protein is utilized when carbohydrate is deficient. This may be compared with the diet of Esquimaux. Carbohydrates are lacking in their environment and are also very bulky foods yielding much less energy, weight for weight, than fats. The almost exclusively meat diet results in a very high protein intake combined with large amounts of fat. This maintains a very high metabolic rate to keep the body warm under Arctic conditions.

BODY TEMPERATURE

The majority of animals are unable to control body temperature, so that it goes up and down with the temperature of the surroundings, i.e. they are **cold-blooded** or **poikilothermic**. Birds and mammals are able to control body temperature to a varying extent, so that it is usually higher than that of the surroundings, i.e. they are **warm-blooded** or **homoiothermic**. (On a hot day in the tropics the temperature of a cold-blooded animal may actually be higher than that of a warm-blooded animal.)

Because warmth activates enzymes, the metabolism of warm-blooded animals usually proceeds at a greater rate than that of cold-blooded animals. Air temperatures vary considerably from day to day and with the time of day, so that warm-bloodedness enables mammals and birds to be active for longer periods than other animals.

Body temperature is maintained by the balance of heat liberated inside the body to heat lost from the body.

Heat is **liberated** during all forms of **activity**. Bulk for bulk, active glandular tissue liberates more heat than any other type of tissue, but in the animal as a whole it is the muscles which are responsible for the greatest amount of heat. Muscular activity is very extravagant, only about 15% of the energy from the oxidized food being available to produce contraction, while the other 85% is in the form of heat.

Heat is **lost** from the **body surface** of cold-blooded animals at such a rate that their temperature is never appreciably higher than that of the surroundings, but warm-blooded animals are all provided with some form of **insulation**. Birds have **feathers** and most mammals have **fur**, while both have large quantities of **subcutaneous fat**. (In man the amount of hair is reduced and clothes are worn instead.)

In spite of this insulation, some heat is lost with the **excretions**, with the **expired air** when the body is warmer than the surrounding air, and from the body surface as a whole. In small mammals, in which the surface area is large compared with the volume, this loss of heat may be considerable. The animal may be unable to maintain the warm-blooded condition in winter and be forced to hibernate. This hibernation is also an adaptation to the period of food shortage, especially for insectivorous and seed-eating types. The rat does not hibernate, but it seldom breeds in winter, so that the young are not exposed to difficult climatic conditions.

The average temperature of the rat is 36·5°C, but the actual temperature is very much more variable than in man. During periods of activity it frequently rises to over 38°C, while during rest in cold weather it may fall to about 32°C without ill effects. The female has a somewhat higher temperature than the male.

In most mammals the body temperature is controlled by varying the amount of blood sent to the skin where it can be cooled, and the amount of sweat produced by the sweat glands, the evaporation of which assists cooling. In the rat the large surface area compared with the volume makes such a cooling mechanism unnecessary and there are no sweat glands in the skin —see page 3. The new-born rat, being hairless, is very susceptible to changes in external temperature.

URINOGENITAL SYSTEM

THE **excretory** and **reproductive systems** of vertebrates are anatomically interrelated. In mammals the relationship is less apparent than in most other groups, and from the functional point of view the systems are separate, except for the use of the urethra of the male as a common urinary and reproductive passage.

(a) EXCRETORY SYSTEM

The excretory system of the mammal consists of two **kidneys**, two **ureters**, a **bladder** and a **urethra**.

KIDNEYS

The kidneys are described as **excretory organs** because they are responsible for the removal of the bulk of the **nitrogenous waste** from the body, but it would be more accurate to describe them as organs of **homeostasis** because they help to maintain the constancy of composition of the **body fluids**. They lie against the dorsal wall of the abdomen, with a thin layer of peritoneum on their ventral sides only. The right kidney of the rat is slightly anterior to the left, while the right kidney of man is slightly inferior to (lower than) the left.

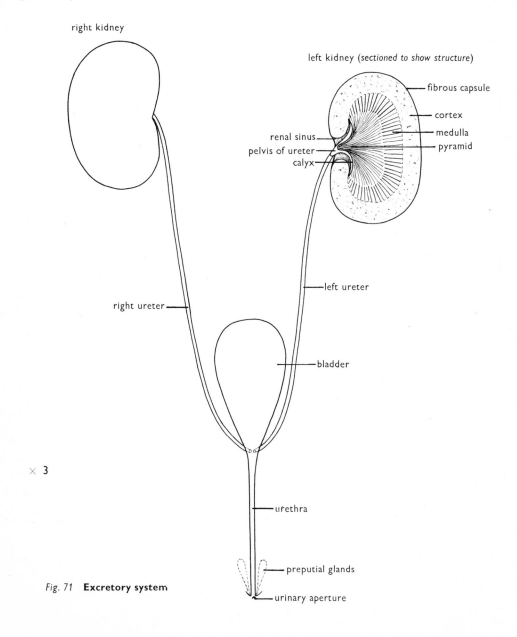

Fig. 71 **Excretory system**

Each kidney is bean-shaped with a notch called the **hilum** where the **renal artery** enters and the **renal vein** and **ureter** emerge. It has a fibrous **capsule** and contains numerous **renal tubules** arranged to form distinct **cortex** and **medulla**. The cortex lies outside the medulla which, in most mammals, is split up into conical sections known as **pyramids**. Each of these lies in a cup-shaped cavity called a **calyx**. The calyces join a wide cavity called the **pelvis** of the kidney, which opens into the ureter and around which is a space traversed by blood-vessels known as the **renal sinus**. The rat has only one pyramid and one calyx, while man has fourteen.

Each **renal tubule** consists of a **Bowman's capsule** and an elongated **duct**. The Bowman's capsule is a microscopic double-walled cup which surrounds a small knot of blood-vessels called a **glomerulus**. Close to each glomerulus are groups of cells forming the **juxtaglomerular apparatus**. The Bowman's capsule and its glomerulus together form a **Malpighian body**. The duct from the Bowman's capsule is differentiated into a **proximal** (first) **convoluted region**, a loop called the **loop of Henle**, and a **distal** (second) **convoluted region**; and it opens into a common **collecting duct** which is shared by several tubules. The Malpighian bodies and convoluted regions of the ducts lie in the cortex and are responsible for its granular appearance, while the loops of Henle and collecting ducts lie in the medulla, making it appear striated. The collecting ducts converge to open at the tip of the pyramid.

The main branches of the **renal artery** anastomose to form ring vessels in the region between the cortex and the medulla. From these vessels smaller branches serve the kidney substance. Each glomerulus has an **afferent vessel** leading to it and an **efferent vessel** leading from it. The latter is narrower than the former and connects with the **capillary network** round the tubules. This network also receives blood directly from the arteries and drains into veins which form ring vessels like the arteries. These veins join to form the **renal vein**. The amount of connective tissue in the kidney is small, so that the blood capillaries come into close contact with the renal tubules and interdiffusion is easy.

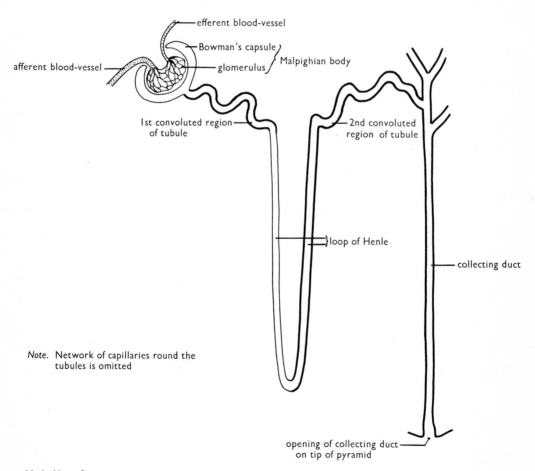

High Magnification

Fig. 72 **A renal tubule (highly diagrammatic)**

FORMATION OF URINE

Most of the experimental work on this subject has been performed on dogs and cats, but the structure of the kidney of the rat is so similar that there is every reason to suppose that the physiology is similar also.

Because the afferent blood vessel to each glomerulus is wider than the efferent vessel, the blood in the **glomeruli** is under considerable **pressure**. This produces **filtration** of fluid from the blood into the **Bowman's capsules**. The filtrate contains **sugar** and **salts** as well as **nitrogenous waste** in solution, but blood proteins and corpuscles are retained in the blood vessels. As this fluid passes down the tubules all the sugar, some salts and some water are reabsorbed and **urine** is formed.

The composition of urine varies with conditions, e.g. the amount of water drunk and the amount of water lost during other physiological processes such as breathing and sweating. Any minute changes in the **osmotic pressure** of the blood are sensed by the **hypothalamus** of the brain which controls the hormonal activity of the **pituitary body**. Water retention is increased by a secretion directly from the pituitary, while sodium retention and potassium loss are promoted by one of the hormones of the adrenal cortex which is itself under pituitary control. The hormones act by affecting the permeability of the walls of the tubules to water and inorganic ions respectively.

Sugar, almost all the sodium, calcium and chloride, and some of the potassium and water to be retained are reabsorbed in the proximal (first) convoluted regions of the tubules. The rest of the chloride and about 8% of the potassium are reabsorbed in the distal (second) convoluted regions. About 20% of the water resorption takes place in the loops of Henle and it is here that the fluid ceases to be isotonic with the blood and the osmotic pressure of the blood around the tubules is adjusted. The pituitary hormone, pitressin, also promotes resorption through the distal convoluted tubules and the collecting ducts. The average composition of human urine is 96% water, 2% salts and 2% urea.

FUNCTIONS OF THE KIDNEYS

As already indicated the kidneys have a variety of functions which can be classified as follows.

(i) **Excretion**. The kidneys cleanse the blood of nitrogenous waste, chiefly in the form of **urea**, and of toxic wastes from certain metabolic processes and also many **drugs**.

(ii) **Homeostasis**. The kidneys regulate the total **volume** of **body fluid**, maintain the **salt and water balance (osmoregulation)** and maintain the **acid/base balance (pH** control).

(iii) **Metabolic activities**. The kidneys can produce **ammonia** and urea from unwanted amino acids, though most of the urea excreted is formed in the liver and brought to the kidneys by the blood. The kidneys can also hydrolyse organic **hexosephosphate** and **detoxicate** many poisonous substances.

(iv) **Secretion**. The kidneys act as endocrine organs. The cells of the **juxtaglomerular apparatus** produce two hormones, **renin** and **erythropoietin**. If blood pressure in the renal artery falls renin is released and converts one of the blood proteins to **angiotensin** which constricts arterioles, thus increasing blood pressure; the increased blood pressure causes renin production to cease, so that the system is self-regulatory. Erythropoietin controls production of erythrocytes by red bone marrow.

URETERS, BLADDER AND URETHRA

Urine passes through the ureters to the bladder, where it is temporarily stored and whence it is passed to the exterior periodically through the urethra.

The **ureters** are long thin passages. They lie against the dorsal wall of the abdomen and open into the back of the bladder close to its neck. The **bladder** is pear-shaped, with a narrow neck which opens into the urethra. Its walls contain a certain amount of muscle and are extremely elastic, so that the capacity can be varied considerably. The lining of the bladder is a special type of epithelium (transitional) which allows for this stretching. The **urethra** of the **male** opens on the tip of the **penis** and is much longer than that of the **female**, which opens immediately anterior to and separate from the opening of the **vagina**. In both sexes there is a **urethral sphincter** of voluntary muscle by which passage of urine (**micturition**) is controlled.

PRO-, MESO-, AND METANEPHRIC KIDNEYS

Kidneys of vertebrates are developed from segmental blocks of tissue called **nephrotomes** which are found in the trunk region only of the embryos. These nephrotomes lie between the dorsal segmental blocks called **myotomes** which give rise to the segmental muscles and the ventral unsegmented **lateral plate mesoderm** which surrounds the coelomic cavity.

The development of the anterior end of the embryo is always slightly in advance of that of the posterior end. The anterior nephrotomes give rise to the first or **pronephric kidneys**. The ducts from these are the **pronephric ducts** and open into the hind end of the alimentary canal which is then known as a **cloaca**.

The more posterior nephrotomes develop slightly later and become the **mesonephric kidneys**, which are the functional excretory organs of adult *fish* and *amphibia*, but are temporary, like the pronephric kidneys, in *reptiles*, *birds* and *mammals*. The mesonephric tubules become connected to the original pronephric ducts so that these then become the **mesonephric** or **Wolffian ducts**. In the majority of **male** vertebrates these ducts are used as **vasa deferentia** for the passage of sperm. In the majority of **female** vertebrates, however, entirely separate ducts, **Mullerian ducts**, are formed from folds of the coelomic epithelium to act as **oviducts**.

In reptiles, birds and mammals the development of the posterior one or two pairs of nephrotomes gives rise to **metanephric kidneys**, the structure of which has been described above. The ducts from these kidneys are entirely separate **metanephric ducts** which are **ureters** only.

(b) REPRODUCTIVE SYSTEM

In the vast majority of vertebrates the **sexes** are **separate**. The **males** have paired **testes** which produce the male gametes or **spermatozoa** and the **females** have one or two **ovaries** which produce the female gametes or **ova**. While all ordinary cell divisions in the body are **mitotic**, i.e. they result in cells of identical nuclear constitution, the final divisions during maturation of gametes are **meiotic** and result in cells with half the normal number of **chromosomes**. The normal number is restored when the gametes fuse at **fertilization**. Most fish and amphibia shed their gametes and fertilization is external but, as an adaptation to life on land, reptiles, birds and mammals have **internal fertilization**. Prior to internal fertilization the male transfers spermatozoa to the ducts of the female during a mating process or **copulation**. Reptiles and birds lay relatively large, hard-shelled eggs, but all mammals except the *Monotremes*, e.g. the duck-billed platypus, are **viviparous**, i.e. they retain the eggs inside the body of the mother so that the young are born in an advanced state of development. In the *Marsupials* or pouched mammals the young are retained for a relatively short time and are born very immature. In the *Eutheria* or placental mammals the young are retained relatively longer and are born in much more mature condition. While in the womb of the mother each **embryo** becomes attached to the mother by a highly vascular structure known as the **placenta**. After establishment of the placenta the embryo is called a **foetus**. In most *Marsupials* the foetal part of the placenta is formed from the **yolk-sac** which is homologous with the sac around the yolk mass in the developing eggs of reptiles and birds, but in a few *Marsupials* and all *Eutheria* the yolk-sac placenta is superseded or replaced by a "true" or **allantoic placenta** as described below.

All mammals possess **mammary glands**. Those of the **males** are **rudimentary** while those of the **females** are active after pregnancy and produce the **milk** on which the young are nourished for some time after birth. Mammary glands usually open on distinct **nipples**. In *Marsupials* these lie inside the pouch so that the very immature young can be fed and sheltered at the same time. In *Eutheria* the nipples are uncovered.

MALE REPRODUCTIVE SYSTEM

The reproductive system of the male mammal consists of two **testes** each with an **epididymis** and a **vas deferens**, two **seminal vesicles** and a number of **associated glands**. The reproductive ducts always join the urethra so that there is a common **urinogenital aperture** at the tip of a **urinogenital papilla** which forms a protrusible **penis**.

In the rat the **testes** are ovoid bodies which develop in the abdominal region but descend to the groin and come to lie in pendulous **scrotal sacs** when the individual is about forty days old. One fold of skin covers both sacs, but they are separate internally and each remains connected to the abdominal cavity by a widely open passage through which the testes are occasionally withdrawn during periods when the sexual function is temporarily in abeyance. The rat has thus no distinct **inguinal canals** such as are found in many mammals, including man. The **interstitial tissue** of the testes produces **testosterone** and other **androgens** responsible for development of **secondary sexual characteristics**, while the tubules produce spermatozoa.

Spermatogenesis, or the production of spermatozoa, takes place continuously throughout adult life. During spermatogenesis, the **germinal epithelium** forming the walls of the **seminiferous tubules** gives off **primary spermatocytes** which undergo reduction division to produce **secondary spermatocytes**. These divide again to form **spermatids** which develop into **spermatozoa**. Each spermatozoon has a **head**, a **middle piece** and a **tail**. The head of the rat spermatozoon is approximately 2·5 μm long. It contains the dense **nucleus** and has a less dense tip called the **acrosome** which helps to penetrate the ovum when fertilization takes place. The tail contains a long **axial filament** which becomes vibratile for a brief period when the spermatozoon is mature. The middle piece contains the **centrioles**, from one of which the axial filament arises, and also a **spirally coiled sheath** of **mitochondrial material** concerned with the respiration necessary to release the energy provided by external nutrients.

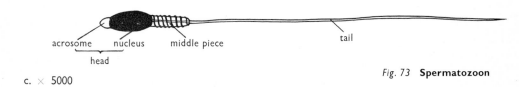

acrosome   nucleus   middle piece   tail

head

c. × 5000

Fig. 73 **Spermatozoon**

Spermatozoa are produced in enormous numbers and pass into the tubules of the **epididymes** where they are stored while they become mature and motile. The epididymes are homologous with the **mesonephric kidneys** of fish and amphibia. Each epididymis is divided into two parts, the **caput epididymis** which lies at the anterior end of the testis and receives ciliated ducts (**vasa efferentia**) from it, and the **cauda epididymis** which lies at the posterior end of the testis and is connected to the caput by a narrow strip. Spermatozoa leave the epididymes through the **vasa deferentia** which are **mesonephric** or **Wolffian ducts**. Each vas deferens loops over the corresponding ureter and opens into the neck of the bladder close to the beginning of the urethra.

Associated with each vas deferens is a **vesicula seminalis**. In the rat these organs are relatively enormous, are curved and sacculated, and are closely connected with two large **coagulating glands**. In man they are very small and in the rabbit they are missing altogether. They produce a viscid secretion which forms much of the volume of the semen.

Like all male mammals, the rat has **prostate** and **Cowper's glands**. The prostate glands are subdivided so that they appear as two pairs of structures, each pair being attached by a stalk to the neck of the bladder. They produce a thin opalescent fluid with a characteristic odour, which aids the motility of the spermatozoa and contributes the enzymes responsible for the coagulation of semen to form the vaginal plug after copulation, and also those responsible for the subsequent liquefaction of the plug. The Cowper's or bulbo-urethral glands open into the urethra at the flexure where it emerges from the pelvis. They produce a mucoid secretion. In addition the rat has a **gland of the vas deferens** lying around the neck of the bladder in the position usually occupied by the prostate glands.

The **seminal fluid** formed by the combined secretion of all these glands bathes the spermatozoa, giving them the nourishment necessary for their motility and viability. It is of sufficient alkalinity to neutralize the acidity of the vagina of the female which would otherwise destroy the spermatozoa rapidly.

The semen is transferred to the vagina of the female during copulation, for which purpose the **penis** is protrusible. A fold of skin at the tip of the penis is called the **prepuce**. The rat has two **preputial glands**.

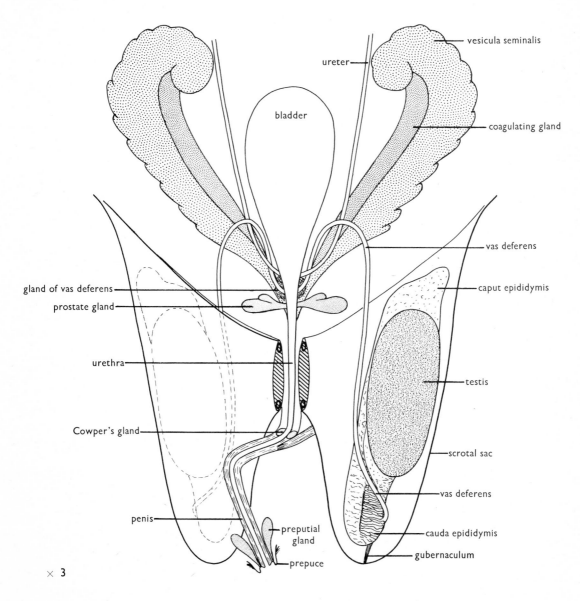

× 3

Fig. 74 **Male reproductive system**

FEMALE REPRODUCTIVE SYSTEM

The reproductive system of the female mammal consists of two **ovaries** and two **oviducts**. The latter are partially joined together and have a common aperture.

The **ovaries** lie in the abdominal cavity, ventro-lateral to the kidneys. Each ovary is small and when mature has a blistered appearance. The ovaries are responsible for the formation of the **ova** and also of the ovarian hormones, the **oestrogens** and **progesterone**, which control development of the **secondary sexual characteristics**.

Most mammalian ova are relatively small with negligible amounts of yolk. The average diameter of the ovum of the rat is 70 μm and of that of man 130 μm. This is in striking contrast to the eggs of *reptiles* (8–40 mm), *birds* (6–85 mm) and *Monotremes* (3–4·5 mm) in which there are large amounts of yolk. The difference can be correlated with the **viviparous** habit and development of the placenta in the *Marsupials* and *Eutheria*.

Oogenesis, or production of ova, starts before birth. In each ovary the **germinal epithelium** is superficial and surrounds a mass called the **stroma**. The germinal epithelium gives off **primary oocytes** and also large numbers of sterile **follicle cells** into the stroma. The follicle cells become grouped round the primary oocytes and multiply to form the **Graafian follicles**. The follicles mature in batches. In the later stages of development they become hollow and contain large amounts of **follicular liquor**, occupying space which would have been filled by the ovum had it been yolky.

The primary oocytes mature to form ova by shedding two minute polar bodies. The **first polar body** is produced by a **reduction division**, and may undergo a further division while the **second polar body** is being given off. The polar bodies then disintegrate. The mammalian ovum is usually ready for fertilization when the first polar body has been produced and the second is about to be formed. At ovulation the ova are shed from the follicles in varying states of maturity, and only those in the correct state are available for fertilization should mating take place.

After ovulation the empty follicles collapse and their cells become converted into yellowish **luteal cells** forming the **corpora lutea**. These degenerate rapidly if pregnancy does not occur, and another group of follicles develops. If pregnancy occurs, the corpora lutea are retained throughout the gestation period. In the rat removal of the corpora lutea even at a late stage terminates pregnancy.

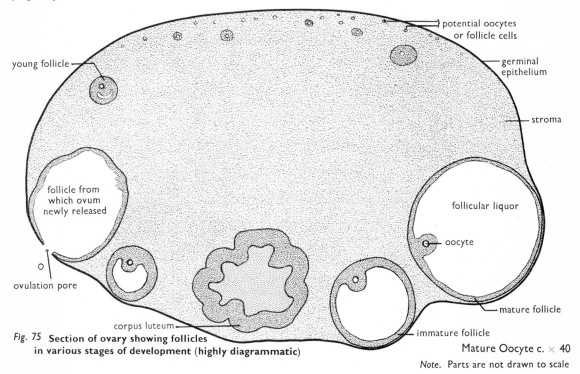

Fig. 75 **Section of ovary showing follicles in various stages of development (highly diagrammatic)**

Mature Oocyte c. × 40

Note. Parts are not drawn to scale

The oviducts are **Müllerian ducts**—see page 53. They are suspended by folds of peritoneum called **mesovaria**. Each oviduct has a narrow region, the **Fallopian tube**, and a wider region, the **uterus**, with a sphincter of muscle at the **tubo-uterine junction**. In the rat the two uteri are sometimes known as uterine horns. They join the **vagina** dorsal to the urethra but ventral to the rectum. The external opening of the vagina is immediately posterior to that of the urethra and a short common passage forms the **vestibule**. Near the opening of the urethra is a small body, the **clitoris**, which is the vestigial homologue of the penis of the male. The whole external genital region forms the **vulva**. In young rats the opening of the vagina is partially covered by a membrane called the **hymen**.

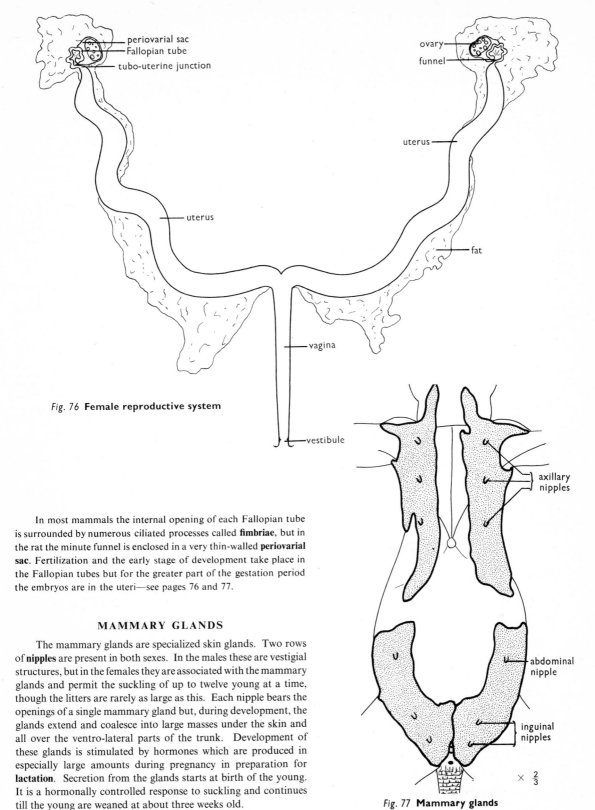

Fig. 76 **Female reproductive system**

Fig. 77 **Mammary glands**

In most mammals the internal opening of each Fallopian tube is surrounded by numerous ciliated processes called **fimbriae**, but in the rat the minute funnel is enclosed in a very thin-walled **periovarial sac**. Fertilization and the early stage of development take place in the Fallopian tubes but for the greater part of the gestation period the embryos are in the uteri—see pages 76 and 77.

MAMMARY GLANDS

The mammary glands are specialized skin glands. Two rows of **nipples** are present in both sexes. In the males these are vestigial structures, but in the females they are associated with the mammary glands and permit the suckling of up to twelve young at a time, though the litters are rarely as large as this. Each nipple bears the openings of a single mammary gland but, during development, the glands extend and coalesce into large masses under the skin and all over the ventro-lateral parts of the trunk. Development of these glands is stimulated by hormones which are produced in especially large amounts during pregnancy in preparation for **lactation**. Secretion from the glands starts at birth of the young. It is a hormonally controlled response to suckling and continues till the young are weaned at about three weeks old.

OESTRUS CYCLE

The interval between successive periods of ovulation or **oestrus** is characteristic of the species concerned. Thus there is a long **anoestrus** period in dogs and cattle but a short **dioestrus** period in cats, rats and man. The **oestrus cycle** is the series of changes which takes place in the ovaries and in the oviducts between one period of oestrus and the next. Unless pregnancy intervenes, the cycle in the rat takes 4–5 days and uterine changes can be recognized by taking a vaginal smear. During dioestrus numerous leucocytes are present, but, as oestrus approaches, the surface cells of the **vaginal mucous membrane** become **cornified** and then **mucilagenous**. In man the uterine mucous membrane breaks down completely about halfway through the dioestrus period and there is bleeding known as **menstruation**. The oestrus (and menstrual) cycles are controlled by **hormones**—see pages 59 and 60. In the rat the regularity of the cycle can be affected by reflex nervous stimulation; e.g. it has been shown that in some cases the amount of light, acting through the eyes, can alter the frequency of oestrus.

External signs of the approach of oestrus in the rat are the swelling of the lips of the vagina and **heat behaviour**. Swelling starts about 36 hours before ovulation. Heat behaviour, i.e. willingness to be mated, precedes ovulation by a comparatively short time (8–20 hours) so that the greatest activity of any spermatozoa transferred at copulation is coincident with the arrival of the ova in the Fallopian tubes. The males are attracted to the females only when the latter are on heat. Following mating the oestrus cycle is inhibited even if the mating is sterile. **Pseudopregnancy** lasts approximately 12 days while **normal pregnancy** lasts 19–21 days.

ENDOCRINE SYSTEM

Glands are of two kinds: (1) **exocrine**, which secrete into cavities and ducts and (2) **endocrine**, which secrete into the blood stream. The mucus, enzymes and other special secretions or excretions from the exocrine glands do not affect living cells or tissues except those of the ducts through which they pass, but the endocrine secretions or **hormones** have remote effects and act as chemical messengers.

The primary action of any hormone is chemical. Hormones affect the permeability of cell membranes and/or induce enzymes possibly by activation of gene sites or combination with repressor molecules. As a result of chemical stimulation physiological changes occur which are secondary effects, but in many cases much more readily observable than the biochemical changes which lie behind them.

The chief endocrine glands are known as **endocrine organs** because the production of hormones is their only recognizable function, but many organs have subsidiary endocrine functions attributable to special groups of cells within them, and there are also general tissue hormones produced widely throughout the body. In health, the entire endocrine system is **self-regulatory**, the presence of one hormone affecting the production of another reciprocally. This feedback mechanism not only produces continuous short-term homeostasis, e.g. regulations of the salt and water content of the blood, but is responsible for the longer-term cyclic events such as the oestrus cycle, and emergency responses such as the mobilization of sugar reserves for "fight or flight".

The endocrine organs of the mammal are the **pituitary body**, the **thyroid gland**, the **parathyroid glands**, the **adrenal glands**, the **thymus** and the **pineal body**. Accessory endocrine function is associated with the **pancreas, kidneys, reproductive organs** and **placenta**, and the **stomach** and **small intestine**.

The **hypothalamus** is not listed as an endocrine organ though it has ultimate control over much of the system through the pituitary body and forms the link with the nervous system. Thus there is integration of neural and hormonal control of body functions.

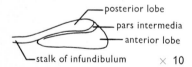

Fig. 78 **Pituitary body (lateral view)**

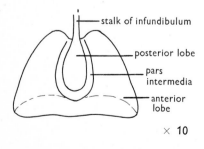

Fig. 79 **Pituitary body (dorsal view)**

PITUITARY BODY

The **pituitary body** or **hypophysis cerebri** is closely associated with the under surface of the brain. Part of it is derived from nervous tissue of the **infundibular region** of the floor of the forebrain and is known as the **neurohypophysis**, while the rest is derived from non-nervous tissue of the roof of the **stomodeal invagination** and is known as the **adenohypophysis**.

Note. In embryological diagrams the rudiment of the adenohypophysis only is usually labelled 'hypophysis', thus producing some confusion in terminology. The evolution of the adenohypophysis is traceable from the nasohypophyseal stalk of jawless fishes and possibly the olfactory pit of *Branchiostoma*.

NEUROHYPOPHYSIS

The **neurohypophysis** consists of the **stalk of the infundibulum** by which the pituitary body is attached to the brain as well as the **posterior lobe** of the pituitary body itself. In the rat the posterior lobe is actually dorsal to the non-nervous part of the gland, but is named by homology with the part which is both dorsal and posterior in man. The posterior lobe contains the end-organs of nerve cells of the hypothalamus. **Neurosecretions** produced in the hypothalamus are transported through the axons and stored till suitable stimulation triggers their release. These secretions include the two posterior lobe hormones—**oxytocin** and **pitressin**—and the **releasing factors** for the hormones of the adenohypophysis.

Oxytocin stimulates contraction of the smooth muscle of the **uterus** during parturition and the **mammary glands** during lactation. **Vasopressin (pitressin)** is primarily an **antidiuretic**, increasing the permeability of the distal convoluted and collecting tubules of the kidneys so that more water is returned to the blood stream and the urine is more concentrated. Injection of vasopressin causes rise in blood pressure by constriction of arterioles and also excites the smooth muscle of the intestine, but it is doubtful if the vasopressor effect is utilized under normal living conditions.

ADENOHYPOPHYSIS

The **adenohypophysis** is formed as an invagination which in many mammals, including man, becomes folded around the neurohypophysis and firmly attached to it, with the original cavity obliterated. In the rat, however, the more primitive condition is found in which the adenohypophysis is easily separable from the neurohypophysis and its cavity, known as **Rathke's pocket**, is retained throughout life. The thin part of the adenohypophysis adjacent to the neurohypophysis is called the **pars intermedia**, while the thicker outside part is called the **anterior lobe**, though in most mammals it lies ventral to the posterior lobe. The secretion from the non-nervous tissue contains a mixture of hormones with widely differing effects. Secretion of the individual hormones is controlled by specific releasing factors of the hypothalamus brought by the portal blood supply. Thus although the adenohypophysis appears to be the "master gland" of the body, it is itself under hormonal control.

The **pars intermedia** produces **melanotropin (MSH)**, so called because it causes dilation of the melanophores of fish and frogs. A vestige of this function is seen in the ability to bring about darkening of mammalian skin, though this effect is normally inhibited by hormones from the adrenal cortex. There is some evidence that melanotropin may help dark adaptation and the resynthesis of visual purple.

The **anterior lobe** produces a number of hormones, the most important of which have been given names corresponding to their principal activity: **growth hormone (GH)**, **thyrotropic hormone (TSH)**, **adrenocorticotropic hormone (ACTH)**, two **gonadotropic hormones (FSH and LH)** and **luteotropic hormone (LTH)**. Some authorities consider that LTH is identical with the growth hormone, but its effects are specific to the adult female.

(a) The **growth hormone (GH)** is also known as **somatotropin (STH)**. It increases general metabolic rate and favours protein synthesis by nitrogen retention, thus promoting growth. The most obvious effect of the hormone is increase in skeletal development. The action is reinforced by the **lipotropic hormone** which stimulates release of fatty acids from adipose tissue and thus makes available reserves of energy-containing substrates. Somatotropin is species specific.

(b) The **thyrotropic hormone (TSH)** stimulates activity of the **thyroid gland** and thus affects the general metabolic rate, but not skeletal development and therefore not growth. Intense emotion and cold increase the production of the appropriate hypothalamic releasing factor and thus of TSH. The consequent increase in thyroxine enhances the effects of adrenaline, and also decreases TSH production by inhibitory feedback.

(c) The **adrenocorticotropic hormone (ACTH)** controls the activity of the **adrenal cortex**, particularly the production of glucocorticoids and sex hormones. Again there is negative feedback with the glucocorticoids inhibiting ACTH production.

(d) The **gonadotropins** and the **luteotropic hormone** are here considered together. In the female the gonadotropins are known as the **follicle stimulating hormone (FSH)** and the **luteinizing hormone (LH)**. In the male, though the name **FSH** is retained, the hormone identical with LH is called the **interstitial cell stimulating hormone (ICSH)**.

The follicle stimulating hormone (FSH) acts on **gonadial cells**. In the female the cyclic predominance of this hormone promotes ripening of the ova in batches, while in the male the more constant level of FSH maintains a continuous supply of spermatozoa throughout adult life.

The luteinizing or interstitial cell stimulating hormone (LH or ICSH) stimulates follicle cells to produce **oestrogen** and the interstitial tissue of the testes to produce **testosterone**. These are closely related substances and are responsible for differentiation of **secondary sexual characteristics**. The oestrogens also bring on **heat** and inhibit FSH release while stimulating increased release of LH. Shortly before ovulation oestrogen production from the follicles ceases, but a small amount is then liberated from the stroma. LH promotes formation of the corpora lutea, but in the rat there is evidence that the final stages of development and assumption of functional activity require the presence of the **luteotropic hormone (LTH)**. The corpora lutea produce **progesterone** which prepares the uterine mucosa for implantation, but alone inhibits LH production. Therefore, if pregnancy does not occur, as the oestrogen production from the ovaries dies away, the corpora lutea atrophy. At the same time, disappearance of the oestrogens permits FSH to be released again and the **oestrus cycle** is complete—see table on page 60.

During **pregnancy** the large amounts of **oestrogens** produced by the placenta are sufficient to maintain high LH release. This LH, reinforced by the LH-like **chorionic gonadotropin**, promotes retention of the corpora lutea and therefore production of the progesterone essential for successful gestation. At the same time release of FSH is continuously inhibited and oestrus is suppressed. The oestrogens and progesterone together stimulate the mammary glands to develop to the pre-lactating condition, but the oestrogens also promote production by the hypothalamus of the prolactin inhibiting factor **(PIF)** which holds back LTH production. With the loss of the placenta at parturition, oestrogens are no longer produced and PIF production ceases. In the absence of inhibition, LTH is secreted and promotes **milk** production from the already prepared mammary glands. As this function of LTH was recognized independently of the luteotropic action, the hormone is also known as the **lactogenic hormone** or **prolactin**. During lactation the luteotropic action is important in promoting the retention of the corpora lutea and progesterone production with consequent suppression of oestrus.

To summarize, the anterior lobe hormones are essential for growth, correct general metabolism, sexual maturity, fertility, and successful breeding. Deficiencies can cause profound behavioural disturbances as well as physiological and morphological abnormalities.

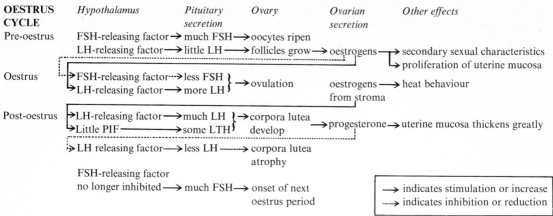

OESTRUS CYCLE	Hypothalamus	Pituitary secretion	Ovary	Ovarian secretion	Other effects
Pre-oestrus	FSH-releasing factor → much FSH → oocytes ripen				
	LH-releasing factor → little LH → follicles grow → oestrogens →				secondary sexual characteristics
					proliferation of uterine mucosa
Oestrus	FSH-releasing factor ····→ less FSH } → ovulation			oestrogens → heat behaviour	
	LH-releasing factor → more LH			from stroma	
Post-oestrus	LH-releasing factor → much LH } → corpora lutea develop			→ progesterone → uterine mucosa thickens greatly	
	Little PIF → some LTH				
	LH releasing factor ····→ less LH ········→ corpora lutea atrophy				
	FSH-releasing factor no longer inhibited → much FSH → onset of next oestrus period				

→ indicates stimulation or increase
····→ indicates inhibition or reduction

THYROID GLAND

The thyroid gland has two **lobes**, one on either side of the anterior end of the trachea, and a band called the **isthmus** which joins the lobes together ventrally. The gland consists of numerous small **vesicles** which produce a group of **iodine-containing** hormones, the best known of which is **thyroxine**. Thyroxine is liberated in response to the presence of the **thyrotropic hormone** (TSH) and controls the general **metabolic rate**, increasing many catabolic processes, particularly the use of carbohydrate. It also increases the rate of heart beat, blood pressure, mental activity, fertility and growth, but decreases the release of TSH from the pituitary and thus regulates its own production. The continual degradation of thyroxine by the liver prevents accumulation of thyroxine and metabolic imbalance. Much of the iodine is recycled.

The thyroid gland also produces **calcitonin** which promotes deposition of **calcium phosphate** in bones and teeth and is therefore antagonistic to parathormone from the parathyroid glands.

In the embryo the thyroid gland develops from a groove in the ventral wall of the pharynx, homologous with the **endostyle** which secretes mucus in primitive **Chordates** such as *Branchiostoma*. It has been shown that the endostylar secretion contains some iodine, so that physiologically as well as anatomically the endostyle is the forerunner of the vertebrate thyroid gland.

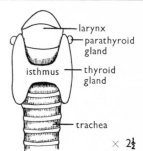

Fig. 80 **Thyroid and parathyroid glands**

× 2½

PARATHYROID GLANDS

The parathyroid glands of mammals are ovoid bodies which lie embedded in the dorsal surface of the thyroid gland. The rat has only two of these glands, while man usually has four. They produce **parathormone**. If blood calcium falls below a critical level tetanic convulsions occur. To prevent this, parathormone mobilizes mineral reserves from bones and teeth and promotes the excretion of excess phosphate by reducing reabsorption of these ions by the renal tubules. Calcitonin from the thyroid gland acts antagonistically to parathormone.

Differences in the calcification of the lines of growth in the incisor teeth of rats may be used as indicators in the biological assay of the potency of parathyroid preparations.

Parathyroid glands develop from the pharyngeal pouches of the embryo and are not found in vertebrates with functional gills, i.e. fishes. They are believed to be homologous in series with the ultimobranchial bodies of the latter, vestiges of which are found in some mammals.

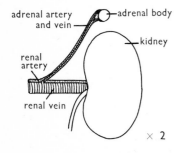

Fig. 81 **Adrenal body**

× 2

ADRENAL GLANDS

Mammals have two adrenal glands (adrenal bodies), one associated with each kidney. In the rat these bodies are rounded and lie slightly anterior to the kidneys, with an independent blood supply. Those of the female are slightly larger than those of the male. The adrenal glands are composed of two types of tissue, **cortex** and **medulla**, with completely separate origins and function.

ADRENAL CORTEX

The adrenal cortex forms a superficial layer in each gland and is derived from the same embryonic tissue as the **kidneys** and **reproductive organs**. It is known to produce over fifty hormones, all of which have been identified as steroids. They may be divided into three groups.

(1) **Mineralocorticoids.** These control mineral salt content of extracellular fluids, particularly the levels of **sodium, potassium** and **chloride**, affecting the permeability of the renal tubules to these ions. Excess of these hormones tends to lead to retention of body fluids and thus oedema.

(2) **Glucocorticoids**. These affect metabolism by increasing **gluconeogenesis** from protein, while inhibiting oxidation of glucose, and thus increasing the amount of blood glucose. They also increase deposition of liver glycogen and total body fat. Besides these general metabolic effects, the glucocorticoids increase **erythrocyte** production, but decrease the number of circulating **lymphocytes**, thus reducing **antibody** production and having an **anti-inflammatory** action. The inhibitory effect on pituitary secretion of ACTH, TSH and MSH provides negative feedback in response to increase in production of the adrenocorticotropic hormone and helps to prevent excessive fluctuation of hormonal levels. Glucocorticoid activity increases in response to stress.

(3) **Sex hormones**. The sex hormones include **androgens** which mimic the effects of the testicular hormones of the male and **oestrogens** and **progesterone** which mimic the effects of the ovarian hormones of the female. Normally production of these hormones is in comparatively small quantities and the hormones from the gonads control development of the secondary sexual characteristics, but it is worthy of note that testosterone is present in the female and can take effect following removal of the ovaries.

ADRENAL MEDULLA

The adrenal medulla lies inside the cortex and is derived from the same tissue as the **sympathetic cords**. It produces **adrenalines** which consist of two hormones, **epinephrine** and **norepinephrine**. As most of the characteristic effects of the medullary secretion are associated with the former, the terms adrenaline and epinephrine are sometimes used synonymously.

Epinephrine increases ACTH and TSH production; breakdown of glycogen in liver and thus output of sugar; breakdown of muscle glycogen and activation of phosphorylase to assist glucose utilization; rate of heart beat and cardiac output; flow of blood to muscles and brain.

Epinephrine decreases flow of blood to the alimentary canal; discharge of erythrocytes from the spleen; time taken for blood clotting.

Norepinephrine has a general **vasoconstrictor** action which, with the increase of cardiac output, raises blood pressure. It also mobilizes **fat** reserves.

Together **epinephrine** and **norepinephrine** aid **neural excitation** of **adrenergic** or **post-ganglionic sympathetic** nerves. The release of adrenaline normally coincides with and reinforces the action of the sympathetic part of the autonomic nervous system —see page 71. This prepares the body for increased physical activity such as is required for "fight or flight" in response to the emotions of fear or anger. Adrenaline is also produced when blood sugar is drastically lowered, and helps to restore the normal level as rapidly as possible.

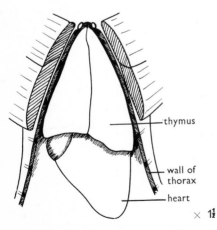

× 1½

Fig. 82 **Thymus gland**

THYMUS

In mammals this gland lies in the thorax anterior to the heart. Like the parathyroid glands, it is derived from the walls of the embryonic **gill pouches**. In the rat it has two distinct lobes and reaches the height of its development at about the 90th day. In man it reaches a corresponding maximum about the time of puberty and disappears altogether in most individuals by the 25th year, its characteristic lymphoidal cells being replaced by adipose tissue. In the rat, however, it is retained throughout life.

Before birth the thymus is a major source of **lymphocytes** and of the precursor cells from which **lymph nodes** and **spleen** develop. The thymus continues to produce lymphocytes throughout life, especially when there has been depletion through stress. Thus it is important in creating and maintaining **immunity**. Two hormones have been isolated from the thymus tissue: **thymosin** which stimulates growth of lymphoid tissue, and **promine** which stimulates general growth. Both hormones are most important in early life when growth is fastest and immunities are being established. Injections of extracts from the thymus have produced precocious development in the rat, but no final increase in growth after the normal adult condition is reached. It has therefore been suggested that the thymus controls rate of attainment of **maturity**.

PINEAL BODY

The pineal body corresponds in position and developmental origin with the **pineal eye** of the most primitive vertebrates. In the mammal it lies between the cerebral hemispheres and dorsal to the corpora quadrigemina—see Figs. 83 and 86. It develops from the roof of the **fore-brain** and, in the rat, becomes globular with a long thin stalk. In the majority of vertebrates the pineal body has no visual function. It degenerates and completely calcifies in many adults, but while of glandular nature produces the hormone **melatonin**, so called because it reduces pigmentation in frogs. In mammals this hormone appears to inhibit secretion of ACTH and the gonadotropic hormones and to have a day/night rhythm which may be significant in establishing the **circadian rhythm** of many physiological functions.

ORGANS WITH ACCESSORY ENDOCRINE FUNCTION

In the **pancreas, kidneys** and **gonads** endocrine function is localized in identifiable groups of cells, which thus act as endocrine organs. In the **stomach** and **small intestine** endocrine function is diffuse and the secretions are sometimes included with the **tissue hormones**. The more typical tissue hormones are, however, produced in a wide range of localities rather than specific organs.

PANCREAS

The endocrine cells form groups known as **islets of Langerhans**, which lie between the exocrine alveoli and produce two hormones—**insulin** and **glucagon**.

Insulin is produced in response to rise in the level of blood sugar especially after meals. Its general effect is to lower blood sugar by promoting uptake into cells. In the absence of insulin, cells are unable to take in glucose and are thus deprived of the basic resource material for energy release. Other substrates can be used temporarily, but ultimately wasting and accumulation of abnormal by-products such as ketone bodies disturb metabolism still further and the animal dies. Meanwhile the most easily observable symptoms of insulin deficiency or *diabetes mellitus* are the rise in the level of blood sugar and the excretion of sugar in the urine.

Besides being essential for uptake of glucose, insulin promotes glycogen formation in muscle and, in conjunction with the growth hormone, increases glucose oxidation. In liver and adipose tissue it promotes the conversion of glucose into fat. In liver it also inhibits release of glucose.

Glucagon raises the level of blood sugar in a similar manner to epinephrine, by stimulating the breakdown of liver glycogen and thus mobilizing reserves. Glucagon and epinephrine are not, however, antagonistic to insulin and the glucose released can be rapidly taken up by other tissues.

KIDNEYS

Close to each glomerulus are groups of cells collectively known as the **juxtaglomerular apparatus**. These produce two hormones—**renin** and **erythropoietin**.

Renin is concerned with maintenance of blood pressure. If the pressure in the renal artery falls renin is released and converts one of the blood proteins to **angiotensin**, which constricts arterioles thus increasing the blood pressure. As a result of this increase the production of renin falls, so that the system is self-regulatory. A second form of angiotensin has adrenal cortex stimulating activity.

Erythropoietin secretion is stimulated by the lack of sufficient oxygen in the blood circulating to the kidneys. It accelerates the formation and maturation of **erythrocytes** and thus increases the oxygen-carrying capacity of the blood.

GONADS

The **ovaries** produce the characteristic female hormones, the **oestrogens** and **progesterone** from the **follicles** and the **corpora lutea** respectively, during the oestrus cycle (see page 60). Testosterone is produced as a transitory stage in the manufacture of one of the oestrogens—oestradiol. Small amounts of oestrogens are produced from cells dispersed in the stroma and are important during the immediate post-oestrus phase. The oestrogens and, to a lesser extent, progesterone are responsible for the development of the **secondary sexual characteristics**. Progesterone is essential for successful **implantation** of the embryo and maintenance of the placenta during gestation. The ovaries are believed to be the source of **relaxin** which is released shortly before parturition to relax and soften the symphysis pubis.

The **testes** produce **testosterone** and a number of other **androgens** similar to those produced by the adrenal glands, but in much greater quantity. These hormones are secreted by the **interstitial cells** under stimulation from ICSH of the pituitary and are responsible for **secondary sexual characteristics**.

PLACENTA

Additional oestrogens and **chorionic gonadotropins** are produced by the established placenta. The chorionic gonadotropins differ chemically from pituitary gonadotropins but resemble the luteinizing hormone (LH) in action, thus increasing oestrogen and progesterone production. The presence of these hormones in urine is the basis of pregnancy tests.

STOMACH

In the presence of certain food substances known as **secretagogues** the cells of the stomach wall produce the hormone **gastrin**, which stimulates the **gastric glands** to produce hydrochloric acid and, when in higher concentration, pepsin.

SMALL INTESTINE

Hormone production in the small intestine is triggered by the arrival of chyme from the stomach. The acidity of the chyme converts the **prosecretin** released to **secretin** which is then absorbed into the blood stream. It stimulates the flow of bile and the secretion of water and bicarbonates from the pancreas so that the acidity is quickly neutralized. A second intestinal hormone known as **cholecystokinin** or **pancreozymin** stimulates contraction of the gall bladder and increases production of enzymes from the pancreas.

TISSUE HORMONES

Tissue hormones are produced widely throughout the body rather than from a localized source. As already mentioned, the hormones from stomach and small intestine—gastrin, secretin and pancreozymin—are sometimes included in this category, but most of the tissue hormones are much less specific in origin and action. In general they are concerned with regulation of **blood pressure**, contraction of **smooth muscle** and **synaptic transfer**.

Blood pressure and local variations in blood distribution are affected by the contraction and dilation of arterioles as a result of changes in the tone of the smooth muscle in their walls.

Vasoconstrictors promote contraction of arterioles. As already described, **norepinephrine** and **vasopressin** have vasoconstrictor action. The chief vasoconstricting tissue hormones are:

(*a*) **angiotensin**, formed from its precursor blood protein by renin from the kidneys;

(*b*) **serotonin**, released from blood platelets during clotting and also responsible for promoting peristalsis of the small intestine;

(*c*) **tyramine**, which stimulates smooth muscle generally.

Vasodilators promote dilation of arterioles by inhibiting the action of the vasoconstrictors. The chief vasodilating tissue hormones are:

(*a*) **bradykinin**, which is formed by the action of pancreatic kallikrein:

(*b*) **histamine**, released in inactive form particularly in the lungs and skin, becomes active when needed to control local circulation, is also responsible for increased secretion of gastric juice, and participates in allergic reactions to produce phenomena like anaphylactic shock:

(*c*) **acetylcholine** (see below);

(*d*) **prostaglandins**, derived from highly unsaturated fatty acids by cyclization and oxidation, particularly abundant in seminal fluid and vasicular glands (whence their name) but found in many other tissues. The prostaglandins stimulate smooth muscle other than that of the arterioles and inhibit fat mobilization from adipose tissue, and are thus antagonistic to norepinephrine.

Hormones acting at synaptic junctions include:

(*a*) **Hydroxytyramine** or **dopamine**, which is the parent substance of epinephrine and norepinephrine and may also be a transfer compound from the ends of sympathetic (adrenergic) nerves;

(*b*) **acetylcholine**, which is a transfer substance for cholinergic nerves and therefore essential neural relay throughout the body with the exception of the post-ganglionic sympathetic termini;

(*c*) γ-**aminobutyrate**, which arises in the brain and appears to block synapses.

Serotonin is also found in the central nervous system where it may have psychic effects.

SUMMARY OF THE EFFECTS OF HORMONES

As can be appreciated from the outline given above, endocrinology is an extremely complex subject. There are innumerable cross-interactions and feedback effects which provide self-regulation while permitting adaptation to varying environmental conditions and emotional stresses.

The overall functions of the hormones are:

(1) maintenance of the **internal environment** within the limits which can be tolerated by the living cells;

(2) control of **metabolism** by regulation of enzyme induction;

(3) **morphogenesis**, i.e. control of development and maturation of gonads and secondary sexual characteristics, and of growth of bones;

(4) integration of **autonomic function** to produce instinctual reactions such as sympathetic responses and sexual and maternal behaviour.

Metabolism is also influenced by availability of metabolites, whether expendable (e.g. glucose) or cyclically reusable (e.g. the ADP $\rightleftharpoons$ ATP reaction), and by the accumulation of waste products of the reactions. The relevant enzymes may be inhibited until the waste products have been dispersed or metabolized. Hormones are not allowed to accumulate, there being enzymes with specific destructive function, e.g. cholinesterase which destroys acetylcholine with extreme rapidity.

NERVOUS SYSTEM

THE rat, like all other vertebrates, has a **hollow dorsal central nervous system**, consisting of **brain** and **spinal cord**, from which **peripheral nerves** supply all parts of the body.

The nervous system is composed of cells called **neurones**, each of which has one or more **processes** radiating from a body or **cyton**, containing a single nucleus. The processes are sometimes naked and sometimes encased in a nucleated sheath, the **neurilemma**, and/or a fatty **medullary sheath**. The cytons are always found in groups. Inside the central nervous system they form the **grey matter** and elsewhere they form masses called **ganglia**.

The neurones are classified according to function. Thus **sensory neurones** transmit impulses from the point of stimulation to the central nervous system and form sensory tracts within the brain and spinal cord. **Motor neurones** form motor tracts within the brain and spinal cord and transmit impulses from the central nervous system to the other parts of the body. Sometimes the impulses are transmitted directly from sensory to motor neurones, but in other cases **association neurones** intervene. The connections between the processes of the neurones are known as **synapses**.

CENTRAL NERVOUS SYSTEM

The **brain** and **spinal cord** develop from the **medullary plate** of the embryo which becomes rolled up into a tube and differentiated. The anterior part of the tube forms **fore-brain**, **mid-brain** and **hind-brain vesicles** from which the corresponding regions of the brain develop later. The posterior part of the tube becomes the **spinal cord**.

Within the brain certain regions of the cavity become dilated to form four **ventricles** while the rest remains as a continuous narrow passage. The first two ventricles are known as the **lateral ventricles**. They communicate with the median **third ventricle** through the **foramina of Monro**. The lateral ventricles and the third ventricle lie in the fore-brain. The **fourth ventricle** lies in the hind-brain and is connected with the third ventricle by the **iter** or **aqueduct of Sylvius**.

The brain and spinal cord are completely ensheathed in three layers of membrane known as the **meninges**. The outermost of these is called the **dura mater** and forms a tough lining for the cranium and vertebral canal. It also supports large venous sinuses within the cranium. The middle layer is the **arachnoid mater**. It is more delicate than the dura mater and is separated from the innermost layer by the **sub-arachnoid space** or **theca**. The innermost of the meninges is the **pia mater**. It is very delicate and highly vascular and follows all the convolutions of the surface of the nervous tissue very closely.

The ventricles and passages of the brain and also the sub-arachnoid space are filled with **cerebro-spinal fluid**, which is a form of lymph produced by two networks of blood-vessels called the **choroid plexuses**. These plexuses form the roofs of the third and fourth ventricles. There is a pore in the roof of the fourth ventricle so that the cerebro-spinal fluid can pass out and bathe the outside, as well as the inside, of the central nervous system. The cerebro-spinal fluid absorbs shock and thus helps to protect the delicate nervous tissue from mechanical damage.

BRAIN

The brain lies within and protected by the cranium from which the cranial nerves emerge through a number of foramina, see page 6.

FORE-BRAIN

The fore-brain or **prosencephalon** is subdivided into the **telencephalon** (anterior) and the **thalamencephalon** (posterior).

A. *TELENCEPHALON*

The telencephalon is divided by a median vertical partition in all vertebrates, so that its cavity forms the two **lateral ventricles** already mentioned. The antero-ventral parts of its walls form the **olfactory lobes**, the **olfactory tracts** and the **hippocampi**. The postero-ventral parts form the **corpora striata** and the dorsal parts form the **pallium**.

The **olfactory lobes** are connected with the olfactory epithelium of the nose by numerous **olfactory nerve** fibres. The **olfactory tracts** connect the olfactory lobes with the **hippocampi**, which are the association centres for the sense of **smell**. The relative size of these parts to the rest of the brain may be taken as an indication of the importance of the sense of smell to the animal concerned. In most mammals, including the rat, they are large, but in man they are very small.

The **corpora striata** are large internal masses of grey matter. They are present in all vertebrates and are connected with instinctive behaviour patterns as opposed to intelligent behaviour. They are extremely well developed in birds but poorly developed in mammals, including the rat.

The **pallium** of lower vertebrates is virtually non-nervous, but in the higher vertebrates it has increasing numbers of nerve cells forming the **cerebral cortex**. In mammals this **neo-pallium** is enlarged to form two relatively enormous **cerebral hemispheres** which contain the highest centres of co-ordination and of intelligence. In many mammals the extent of the cerebral cortex is increased by **convolutions** of the surfaces of the hemispheres. There are few such convolutions in the rat but many in man, and the general levels of intelligence of these species can be correlated with the ratio of the total area of the cerebral cortex to the bulk of the individual in each case. In man the cerebral cortex has been mapped, i.e. the various regions have been shown to be concerned with special sensory or motor functions. The cerebral cortex of other mammals could be similarly mapped.

In all *Eutheria* or placental mammals the cerebral hemispheres are linked to one another by a band of fibres known as the **corpus callosum** which is bent back on itself at the **splenium**. This connection adds to the efficiency of the co-ordination of the two sides of the brain.

B. *THALAMENCEPHALON*

In mammals the **thalamencephalon** or **diencephalon** is completely overhung and masked in dorsal view by the cerebral hemispheres. The floor of this region of the embryonic fore-brain gives rise to the **optic vesicles, hypothalamus** and **infundibulum**, the side walls to the **thalami** and the roof to the **anterior choroid plexus** and the **pineal body**. The cavity of the thalamencephalon is the **third ventricle**.

The **optic vesicles** become the **retinae** of the eyes, see page 71. Fibres from the ganglion cells of the retinae grow down the **optic stalks** to form the **optic nerves**. A number of these fibres cross to the opposite side of the brain and thus form the **optic chiasma**. The optic nerve fibres end in the posterior parts of the **thalami**, which thus bear the same relationship to the optic sense as the hippocampi to the olfactory sense.

The **thalami** are thick and meet one another across the cavity of the third ventricle as the **massa intermedia**. They are nerve centres where almost all the sensory impulses reaching the fore-brain are relayed before transmission to the cerebral cortex.

The **infundibulum** becomes the neurohypophysis or neural-derived section of the pituitary body. The **neurohypophysis** consists of the infundibular stalk containing axons and the posterior lobe containing end-organs of nerve cells of the hypothalamus. The latter is the main link between the nervous system and the endocrine system—see page 58.

The **anterior choroid plexus** is non-nervous and consists almost entirely of a dense network of blood-vessels.

The **pineal body** is attached to the brain by the **pineal stalk**. It is the vestigial remains of the third eye possessed by certain primitive vertebrates and is believed to have an endocrine function, see page 60. The pineal body of the rat is rounded and is visible in a dorsal view of the brain, lying between the cerebral hemispheres and the cerebellum. The pineal stalk is long and slender.

MID-BRAIN

The mid-brain or **mesencephalon** consists of the **optic lobes** and the **crura cerebri**. Its cavity is the narrow canal called the **iter** which connects the third ventricle with the fourth ventricle.

The **optic lobes** are dorsal. Those of mammals are subdivided to form four **corpora quadrigemina**. Only the anterior corpora quadrigemina have optic connections, while the posterior pair are concerned with relay of auditory sensation. This arrangement is correlated with the great elaboration of the mechanism of hearing in mammals.

The **crura cerebri** are thick bands of nerve fibres on the floor of the mid-brain. They consist of tracts linking the fore-brain with the hind-brain and spinal cord.

HIND-BRAIN

The hind-brain or **rhombencephalon** consists of the **cerebellum**, the **pons Varolii** and the **medulla oblongata**. Its cavity is the **fourth ventricle**, the roof of which is the **posterior choroid plexus**.

The **cerebellum** is dorsal. Like the cerebral hemispheres, it has superficial grey matter forming **cortex**, the area of which is increased by **convolutions**. In mammals in general the median portion of the cerebellum is called the **vermis** and the lateral extensions are the **flocculi** and **paraflocculi**. In man these terms are not used because the two halves of the cerebellum form distinct **hemispheres** without lateral extensions. Inside the cerebellum of mammals numerous tracts of nerve fibres form the branching **arbor vitae**.

The cerebellum is directly connected to the organs of **balance** in the ears and contains the centres which co-ordinate **muscular activity** and thus maintain **posture**. **Conditioned reflexes** (see page 69), are also due to the establishment of connections in the cerebellum.

The **pons Varolii** is a bridge of nerve matter on the ventral side of the mammalian brain and largely masks the crura cerebri from external view. It contains **relay centres** from the cerebral cortex to the cerebellum and also centres of origin of some of the cranial nerves.

The **medulla oblongata, metencephalon** or **myelencephalon** is like an enlarged region of the spinal cord and is sometimes called the **spinal bulb**. It contains the centres of origin of several of the **cranial nerves** and also the **special centres** which govern involuntary activities such as the rate of **heart-beat, swallowing, vomiting** and **breathing**.

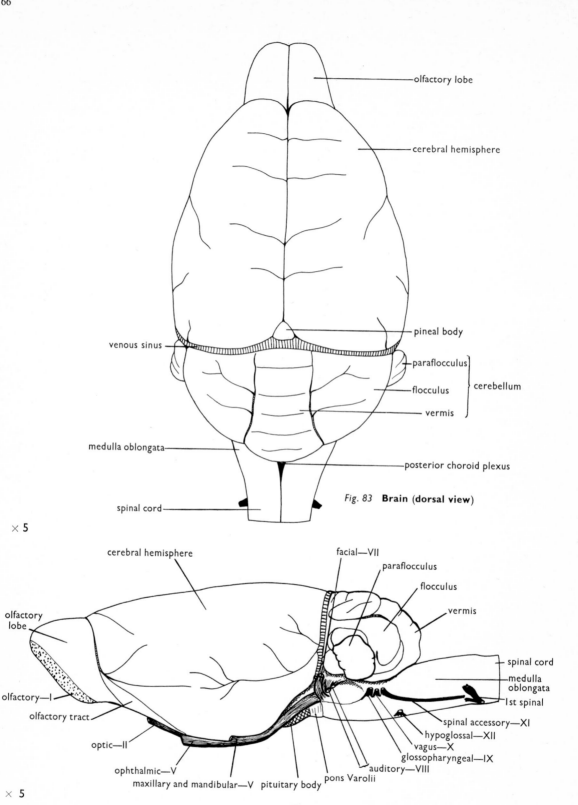

olfactory lobe

cerebral hemisphere

pineal body

venous sinus

paraflocculus

flocculus — cerebellum

vermis

medulla oblongata

posterior choroid plexus

Fig. 83 **Brain (dorsal view)**

spinal cord

× 5

cerebral hemisphere

facial—VII

paraflocculus

flocculus

vermis

olfactory lobe

spinal cord

medulla oblongata

1st spinal

olfactory—I

olfactory tract

spinal accessory—XI

hypoglossal—XII

vagus—X

glossopharyngeal—IX

optic—II

auditory—VIII

ophthalmic—V

pons Varolii

maxillary and mandibular—V

pituitary body

× 5

Fig. 84 **Brain (lateral view)**

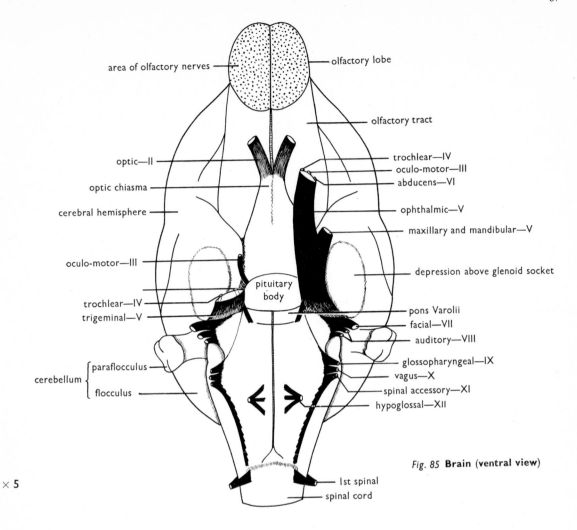

area of olfactory nerves

olfactory lobe

olfactory tract

optic—II

trochlear—IV
oculo-motor—III
abducens—VI

optic chiasma

ophthalmic—V

cerebral hemisphere

maxillary and mandibular—V

oculo-motor—III

depression above glenoid socket

pituitary body

trochlear—IV
trigeminal—V

pons Varolii
facial—VII
auditory—VIII

cerebellum { paraflocculus
flocculus

glossopharyngeal—IX
vagus—X
spinal accessory—XI
hypoglossal—XII

1st spinal
spinal cord

Fig. 85 Brain (ventral view)

× 5

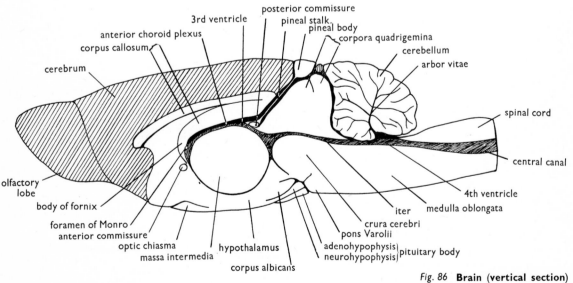

posterior commissure

3rd ventricle
pineal stalk
pineal body

anterior choroid plexus
corpus callosum

corpora quadrigemina
cerebellum
arbor vitae

cerebrum

spinal cord

central canal

olfactory lobe

body of fornix

4th ventricle

foramen of Monro
anterior commissure

iter
medulla oblongata

optic chiasma
massa intermedia

hypothalamus

crura cerebri
pons Varolii
adenohypophysis } pituitary body
neurohypophysis

corpus albicans

Fig. 86 Brain (vertical section)

CRANIAL NERVES

There are **twelve pairs** of cranial nerves in mammals. Some of these are entirely **sensory**, some entirely **motor** and the rest **mixed**, including both sensory and motor fibres.

No.	Name	Type	Distribution
I	Olfactory	sensory	Olfactory epithelium of the upper parts of the nasal cavities
II	Optic	sensory	Retinae of the eyes. Many fibres cross in the optic chiasma
III	Oculomotor	motor	Superior, inferior and internal (anterior) rectus muscles and inferior oblique muscles of the eyes
IV	Trochlear	motor	Superior oblique muscles of the eyes
V	Trigeminal	mixed	Ophthalmic branch sensory from snout Maxillary branch sensory from the upper jaw region Mandibular branch sensory from lower jaw region and motor to the muscles of mastication
VI	Abducens	motor	External (posterior) rectus muscles of the eye
VII	Facial	mixed	Numerous branches which are sensory from the upper parts of the face, the back of the head, the jowl region and the palate, and are motor to the muscles of the face
VIII	Auditory	sensory	Cochlear branch from the cochlea (auditory region of the ear) Vestibular branches from the vestibule and ampullae (balance regions of the ear)
IX	Glosso-pharyngeal	mixed	Tongue (including taste buds) and pharynx
X	Vagus	mixed	Mainly autonomic fibres to the viscera, including the heart and great blood-vessels, the lungs and respiratory passages, the alimentary canal (except the rectum), the liver, pancreas and spleen
XI	Spinal accessory	motor	Some of the muscles of the neck
XII	Hypoglossal	motor	Muscles of the tongue and hyoid region

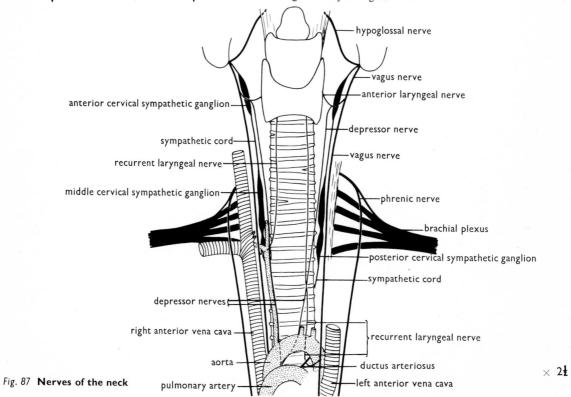

Fig. 87 **Nerves of the neck**

× 2½

SPINAL CORD

The spinal cord lies in the canal formed by the neural arches of the vertebrae. It is cylindrical with two deep longitudinal grooves, the **dorsal** and **ventral fissures**. The diameter of the nervous part of the spinal cord is uniform except for a slight thickening at the base of the neck and another in the lumbar region, where the nerves to the limbs are given off. At its posterior end the spinal cord tapers sharply to a non-nervous thread called the **filum terminale**. In the rat the nervous part of the cord extends about two-thirds of the length of the vertebral canal and the filum terminale starts opposite the second lumbar vertebra.

The spinal cord is traversed by a very narrow **central canal** which is continuous with the cavity of the fourth ventricle. In transverse sections of the cord this canal is seen to lie close to the inner end of the dorsal fissure. The **grey matter** is arranged in an approximately H-shaped pattern with **dorsal** and **ventral horns** and is surrounded by **white matter** consisting of tracts of medullated nerve fibres.

The spinal cord relays impulses in and out at the same level, up and down to different levels, and to and from the brain.

SPINAL NERVES

The spinal nerves are all **mixed nerves** with **sensory** and **motor fibres**, but each of them has **two roots** in which these fibres are segregated. The sensory fibres form the **sensory** or **dorsal roots** of the nerves on the course of each of which there is a **dorsal root ganglion**. The motor fibres form the **motor** or **ventral roots** of the nerves on which there are no ganglia. The roots join just before the nerves emerge from the vertebral canal through the **intervertebral notches**. Many of the nerves emerge considerably posterior to their points of origin from the spinal cord and those which lie parallel to the filum terminale form the **cauda equina**.

Almost all the spinal nerves give off small **dorsal branches** to the skin and muscles of the back, while the **main branches** serve the lateral and ventral parts and the limbs. In the regions of the limbs these ventral branches form networks or **plexuses**.

The rat has thirty-four pairs of spinal nerves: 8 cervical, 13 thoracic, 6 lumbar, 4 sacral and 3 caudal. Each **brachial plexus** is composed of six nerves (the 4th cervical–1st thoracic inclusive) and serves the shoulder region and fore-limb. Each **lumbo-sacral plexus** is composed of seven nerves (the 1st lumbar–1st sacral inclusive) and serves the hip region and hind-limb. There is also a small **pudendal plexus** on each side consisting of branches from seven nerves (6th lumbar–2nd caudal inclusive) and serving the region round the anus and the muscles of the tail.

Branches of the 5th cervical nerves, with occasionally branches from the 4th cervical nerves also, become the **phrenic nerves** to the diaphragm.

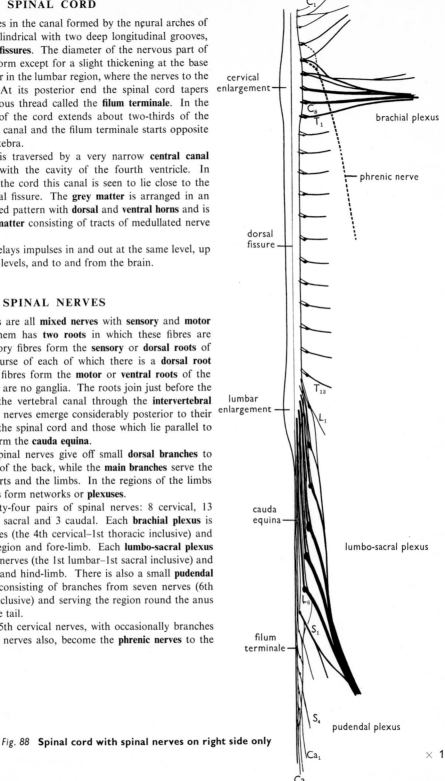

Fig. 88 **Spinal cord with spinal nerves on right side only**

× 1

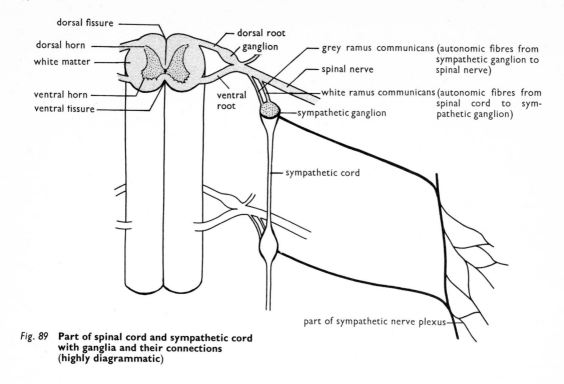

Fig. 89 **Part of spinal cord and sympathetic cord with ganglia and their connections (highly diagrammatic)**

AUTONOMIC NERVOUS SYSTEM

This term is applied to all those parts of the nervous system which control **involuntary activity**. The autonomic system can be divided into the **parasympathetic system** and the **sympathetic system**. Each of these systems consists of **preganglionic autonomic nerve fibres** from the central nervous system to **ganglia** outside it and **postganglionic autonomic nerve fibres** from these ganglia to the **involuntary muscle** and **glands**, most of which are doubly innervated, i.e. they are served by parasympathetic and sympathetic fibres. The preganglionic fibres are **medullated** but the postganglionic fibres are **non-medullated**.

A. *PARASYMPATHETIC SYSTEM*

The **preganglionic parasympathetic fibres** are relatively long. They leave the brain in the **third, seventh, ninth** and **tenth cranial nerves** and the spinal cord in some of the **sacral nerves**. The autonomic fibres in the oculomotor nerves go to the **iris** and **ciliary body** of the corresponding eye. Those in the facial and glossopharyngeal nerves go to the **salivary glands**, while those in the vagus nerves are far more numerous than the non-autonomic fibres and go to the **heart** and **great blood-vessels**, the **lungs** and **respiratory passages**, the **alimentary canal** (except the rectum), the **liver**, the **pancreas** and the **spleen**. The sacral autonomic fibres form the **pelvic nerves** which go to the remaining viscera, i.e. the **kidneys**, the **bladder**, the **reproductive organs** and the **rectum**.

The **parasympathetic ganglia** lie in the organs served.

The **postganglionic parasympathetic fibres** are relatively short.

B. *SYMPATHETIC SYSTEM*

The **preganglionic sympathetic fibres** are relatively short. They leave the spinal cord with the **thoracic** and **lumbar nerves**, which they accompany for a short distance only before passing through **white rami communicantes** to the sympathetic ganglia.

The **sympathetic ganglia** are joined together by strands containing some pre- and some postganglionic fibres and form two **sympathetic cords** close to the vertebral column. In the thoracic and lumbar regions the sympathetic ganglia correspond to the spinal nerves, to which they are connected by the white rami communicantes, but there are additional sympathetic ganglia in the cervical and sacral regions.

The **postganglionic sympathetic fibres** are relatively long. Some of them pass through **grey rami communicantes** to join spinal nerves, while others form separate **sympathetic nerves**.

The rat has 3 cervical, 10 thoracic, 6 lumbar, 4 sacral and 1 caudal pairs of sympathetic ganglia. The anterior cervical ganglia lie close to the angle of the jaw. The middle and posterior cervical ganglia are at the base of the neck, and the short stretch of cord between them is divided to form a loop round the subclavian artery. The 1st thoracic to 1st sacral ganglia are connected to the corresponding spinal nerves by white and grey rami, but the middle cervical ganglia are connected to the 5th to 8th cervical nerves by grey rami only. Fibres passing through the grey rami accompany the spinal nerves to their distribution in skeletal muscles and skin. Additional fibres from the 8th thoracic to 1st lumbar ganglia join to form the **greater splanchnic nerves** which serve the **cardiac** and **coeliac ganglia** near the roots of the coeliac and anterior mesenteric arteries. From these ganglia sympathetic fibres follow the branches of the arteries to the adjacent viscera. The ganglia and the network of fibres associated with them constitute the **solar plexus**. Similarly fibres from the 3rd and 4th lumbar ganglia form the **lesser** and **least splanchnic nerves** which join the **inferior mesenteric plexus** and serve the posterior viscera. Sympathetic nerves from the superior cervical ganglia join cranial nerves to serve the **iris** and **ciliary** body of the corresponding eye and the **salivary glands**, while those from the inferior cervical ganglia go directly to the **heart**.

EFFECTS OF THE AUTONOMIC SYSTEM

The parasympathetic and sympathetic components of the autonomic system act **antagonistically**. The balance of their action produces smooth changes in tone of **smooth muscle** and in the activity of **glands** in contrast to the more abrupt changes shown by skeletal muscle.

Some of the effects of the autonomic system are shown in the following table:

PARASYMPATHETIC	SYMPATHETIC
Contracts ciliary muscles—accommodation for near vision	Relaxes ciliary muscles—accommodation for distant vision
Contracts circular muscles of iris—reduction in size of pupils	Contracts radial muscles of iris—increase in size of pupils
Increases secretion of glands	Decreases secretion of glands
Slows and weakens heart-beat	Quickens and strengthens heart-beat
Contracts bronchioles	Dilates bronchioles
Increases peristalsis of alimentary canal	Decreases peristalsis of alimentary canal
Relaxes sphincters of alimentary canal	Constricts sphincters of alimentary canal

THE METHOD OF FUNCTIONING OF THE NERVOUS SYSTEM

EXCITATION AND CONDUCTION

The **excitability** and **conductivity** of nerve cells are responsible for the reception and transmission of information from one part of the body to another with extreme rapidity.

In the resting neurone the **semipermeability** of the living cell membrane permits the establishment of inequalities in ion distribution between the intracellular fluid and the interstitial fluid. The tendency to equalize along the diffusion gradient is still there, but the accumulation of organic anions (A^-) inside the cell repulses chloride ions (Cl^-) while attracting the cations of potassium (K^+) and sodium (Na^+). Both cations can pass through the cell membrane but permeability to K^+ is consistently high, while that to Na^+ is very low. Equilibrium distribution of K^+ and Na^+ is never reached because active transfer by the **sodium pump** (strictly the Na^+–K^+ pump) transports Na^+ ions out of the cell while slightly increasing the concentration of K^+ inside the cell. The energy required is released from ATP.

The result of the semipermeability and the action of the sodium pump is the establishment of a **voltage difference** of 60–80 mV across the cell membrane. By convention $+$ and $-$ signs are used to indicate direction of **polarization**. Thus -70 mV **resting potential** indicates that the membrane is 70 mV negative inside to outside.

Nerve and muscle fibre membranes can, when suitably stimulated, undergo transient permeability changes during which Na^+ ions can enter the cell rapidly and the voltage across the membrane is immediately decreased. Depolarization of about 10 mV increases permeability to Na^+ ions, causing further voltage drop which in turn gives even greater Na^+ permeability. The depolarization or action potential at the initial input point keys the depolarization of adjacent membrane, thus the action potential spreads as a **self-propagating wave**.

Unchecked, the membrane potential would swing to $+60$ mV (the equilibrium for Na^+), but the depolarization is transient and permeability to Na^+ falls to the resting value in about a millisecond. At the same time permeability to K^+ increases so that the overshoot is limited to 10–30 mV. During repolarization there is a brief hyperpolarization to -75 mV before the resting state of -70 mV is restored after about 100 milliseconds.

The **stimulus** to initial depolarization may be **light**, **heat**, **mechanical distortion** or **chemicals**, but not, except during accidental or experimental circumstances, electrical. A threshold value known as the receptor or generator potential is needed to trigger the self-propagating action potential, which then passes along the fibre without decrement. Thus though the stimulus may be graded the response is all or nothing. Increased activity only occurs if more fibres are involved.

In myelinated nerve fibres impulse propagation is from node to node, so conduction speed is increased without increase in diameter of the axon; and it is possible for vertebrates to have very large numbers of sensory fibres and somato-motor fibres, with much greater flexibility of reaction pattern than the invertebrates with their dependence on giant fibres for rapid conduction of impulses.

INTERNEURONAL AND NEUROMUSCULAR TRANSMISSION

The nervous system is built up of a complex pattern of neurones with interconnections known as **synapses**. Transmission across **synaptic junctions** depends on the formation by **excitatory synaptic knobs** of a variety of chemical transmitter substances of which the commonest is acetylcholine. The transmitter substances are released by arrival of the wave of action potential at the nerve termini and affect the adjacent **postsynaptic membranes** so that the latter become highly permeable to all ions and thus initiate action potential in the **postsynaptic neurones**. Similarly, chemical transmission is responsible for release of action potential in muscle fibres and thus initiation of muscle contraction.

Skeletal muscle is not spontaneously active and depends on liberation of acetylcholine at the motor end-plates on the individual fibres. Visceral smooth muscle is spontaneously active and is stimulated by stretch and by hormones as well as being controlled by the amount and kind of motor input. The **cholinergic** (parasympathetic) autonomic fibres liberate **acetylcholine** while the **adrenergic** (sympathetic) fibres liberate **epinephrine** and **norepinephrine**.

Synaptic knobs can also produce substances capable of increasing the membrane potential of postsynaptic neurones, thus having **inhibitory** action.

THE FUNCTIONS OF THE NERVOUS SYSTEM

Conduction in axons and at synapses is **unidirectional**. Therefore the nervous impulses are relayed from the sensory input to the effector output. The essential components of the relay system are known as the **reflex arc**. Reflex arcs vary in complexity, but the basic parts are:

(1) a **sensory receptor** which may be a naked nerve ending or a specialized sensory cell capable of reacting to (*a*) stimulation from outside the body (**exteroceptor**) or (*b*) internal stimulation (**proprioceptor** or **enteroceptor**);

(2) an afferent or **sensory neurone** whose cell body is in one of the dorsal root ganglia or the homologous ganglia of the sensory cranial nerves;

(3) a **relay centre** within the central nervous system which may consist of synaptic junctions only in the monosynaptic reflex arcs, or involve one or more interneurones in multisynaptic arcs;

(4) an efferent or **motor neurone** whose cell body is in the grey matter of the central nervous system;

(5) an **effector** structure which may be **muscle** or **gland**.

REFLEX ACTION

A **reflex action** is the response brought about by impulses passing through a reflex arc. Simple or **unconditioned reflex** action is unconsidered and is invariable for a given stimulus. Unconditioned reflexes are inherent, i.e. they develop as a result of heredity rather than as an effect of the environment. Somatic reflexes utilizing skeletal muscles enable the organism to react to its environment as perceived by the exteroceptors, which are dealt with in the next section. When the motoneurones supplying a given muscle are reflexly excited by an afferent volley, those supplying the antagonistic muscles are invariably inhibited. Monosynaptic reflexes always discharge to the muscle in which the stimulus originates, which with the usual reciprocal inhibition enables an action, once started, to be continued without further external stimulation. The sense organs within the muscle are proprioceptors. Multisynaptic reflexes typically involve several sections of the spinal cord and/or the brain and result in much more complex reflex action patterns. These reflexes can be blocked by conscious thought and are therefore described as **voluntary**. The inhibitory volleys act by blocking the excitatory impulses in the presynaptic fibres. **Involuntary** reflexes which affect visceral muscles and glands cannot be blocked deliberately. They regulate the functioning of visceral organs and adjust the internal environment to match external changes and behaviour. Enteroceptors within the viscera are linked to the autonomic nervous system, but exteroceptors may be involved, e.g. smell and the secretion of saliva. Both voluntary and involuntary reflexes can become conditioned by experience.

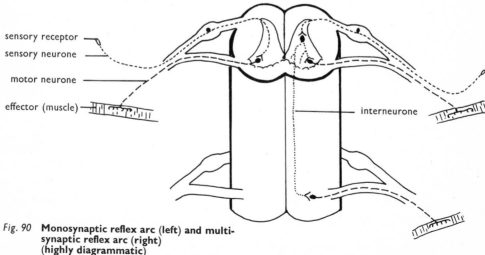

sensory receptor
sensory neurone
motor neurone
effector (muscle)
interneurone

Fig. 90 **Monosynaptic reflex arc (left) and multi-synaptic reflex arc (right) (highly diagrammatic)**

EXAMPLES OF REFLEX ACTION

(1) Whenever a rat is disturbed or startled it makes avoiding movements. The stimulus may be appreciated by any of the senses and the response is made by voluntary muscles over which higher control is possible, should there be warning in advance of the disturbance and should experience and intelligence show lack of response to be desirable.

(2) Whenever anything is passed rapidly in front of the eyes, blinking of the eyelids takes place. The stimulus is appreciated by the eyes and the response is made by the voluntary muscles in the eyelids. This reflex action is designed to protect the surfaces of the eyeballs against damage.

(3) Accommodation (focusing of the eyes) and control of the size of the pupils are both reflex actions. The stimuli are the clarity of the image and the amount of light entering the eye respectively. The responses are by involuntary muscles and therefore cannot be affected by thought. Both these responses are brought about by changes in the balance of para-sympathetic and sympathetic stimulation.

(4) Salivation is also a reflex action. The stimulus may be the movement of chewing or the smell of food. Experiments on dogs have shown that salivation can become a conditioned reflex and occur in response to other regularly occurring stimuli such as a bell at meal-times. Production of saliva is, however, always an involuntary activity under the control of para-sympathetic and sympathetic nerve fibres.

INSTINCT

It is difficult to distinguish between instinct and learning in mammals. Experiments with rats have shown that hoarding, nest making, mating and maternal behaviour appear to be fundamentally instinctive but even these may be modified by circumstances. Thus, e.g. if rats are isolated from birth by Caesarian operation (Kaspar Hauser treatment) and deprived of all transportable material, they fail to make satisfactory nests when material is made available to them at breeding time. On the other hand the consummation of an instinctive action appears to provide satisfaction, irrespective of results: e.g. young which have been fed abnormally fast will go on sucking extraneous objects until the sucking urge is satisfied. On the whole instinctive behaviour is such as to adapt the rat to circumstances it would normally encounter in the wild environment; e.g. it hoards food when a plentiful supply is available, thus providing for periods of shortage.

LEARNING

Learning is the modification of behaviour by experience. All learning is prompted by **motivation** such as hunger, fear, anger or the sexual urge, but the learning itself is of various types.

(1) **Habituation** is the waning of response to stimulation which has been repeated frequently without producing any appropriate result or satisfaction of the motivation. It thus differs from the waning of response through fatigue, age or state of maturity, or injury.

(2) **Conditioning** is the association between a secondary stimulus and a primary one, so that the secondary stimulus by itself becomes capable of eliciting the reflex action, e.g. the salivation at the sound of a bell mentioned above. Conditioned reflexes are subject to habituation and therefore fade or are inhibited if the motivation ceases to be satisfied, but may revive after a lapse of time indicating that there is memory, the strength of which is dependent on the frequency and recency of the conditioning.

(3) **Training** is a form of learning which depends on a general variability of behaviour. The desired result is at first achieved purely by **trial and error** but errors are gradually eliminated till a successful pattern of action is built up. This pattern remains established so long as motivation is satisfied, but shows gradual inhibition if satisfaction is denied. Effort continues longer, i.e. inhibition is slower, if the reward was originally given irregularly. Adaptability to training is dependent on the observation, memory and intelligence of the individual. The rate of improvement is also affected by **practice**, too much causing mistakes to be repeated, too little causing them to be forgotten.

Numerous experiments have shown that rats can learn to find their way through mazes, to tap levers or to push trap doors and to associate these activities with noises, lights or selection of simple symbols. The motivation is usually hunger and the reward food, but dislike and avoidance of electric shock have also been used.

INTELLIGENCE

Intelligence is the ability to work out a problem in the absence of direct experience. The amount by which intelligence depends on heredity or on experience is still under discussion. There is no doubt that certain hereditary chemical deficiencies lead to failure to make intelligent responses or respond to training, but there is also considerable evidence that stimulating variations in the surroundings of the young, promoting interest rather than boredom, both improve memory and increase deductive and intelligent behaviour.

In rats intelligent behaviour has been recorded in the methods used for removing eggs. They have been seen to roll eggs with their noses, and also to co-operate in the transport of eggs over difficult places by one rat lying on its back, holding the egg with its four legs while the other drags it along by its tail or by a straw held in its mouth.

SENSITIVITY

The body is sensitive to a wide range of different types of stimuli.

TOUCH

The sense of touch is distributed all over the body and there are touch fibres attached to each of the hair follicles so that movements of the hairs can be felt. The **vibrissae** or whiskers are especially long and sensitive hairs and are used by the rat to avoid obstacles in the dark.

CHEMICAL SENSE

Taste and **smell** are both chemical senses. Taste in the mammal is the appreciation of certain types of substance in solution, while smell is the appreciation of volatile substances.

Taste is appreciated by **taste buds** in the mucous membrane of the tongue and palate. The distribution of these varies in different mammals. Each taste bud consists of a group of sensory cells surrounded by supporting cells. Each **sensory cell** has a minute projecting **sensory process** by means of which the stimulus is perceived. The impulse is relayed to a sensory nerve fibre, the free end of which is wrapped round the sensory cell. Each taste bud can react only to a restricted number of dissolved substances giving the sensation of saltiness, sweetness,

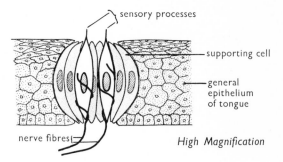

Fig. 91 **Taste bud showing two sensory cells (highly diagrammatic)**

acidity and bitterness according to the type of bud. Though some substances may also produce a feeling of burning on the tongue, the majority of flavours are really smelt.

Smell is appreciated by the **olfactory epithelium** of the upper regions of the nasal cavities. Here, amongst the cells of the mucous membrane, there are numerous naked nerve endings which are stimulated directly. The sense of smell is extremely well developed in most mammals, though feebly developed in man. The rat makes considerable use of this sense when searching for food at night.

SIGHT

The organs of sight are the **eyes**. The vertebrate eye consists of an **eyeball** held in place in its **orbit** by **six muscles** and usually capable of being covered over and protected by **eyelids**.

In mammals the four **rectus muscles** lie at the back of the eye and can move it up and down or from side to side. The **oblique muscles** are attached approximately to the equator of the eyeball. The inferior oblique originates directly from the floor of the orbit, while the superior oblique arises from the back and has its course diverted by passing through a tendinous loop above the upper eyelid. The oblique muscles balance the tendency of the recti to twist the eyeball.

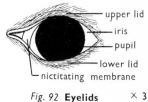

Fig. 92 **Eyelids** × 3

Between the eyeball and the wall of the orbit most mammals have a pad of fat, but the rat has exceptionally large **intraorbital lacrimal glands** and large **Harderian glands** which fill this space completely, see Fig. 54.

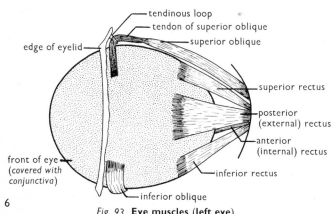

× 6

Fig. 93 **Eye muscles (left eye)**

The majority of mammals, including the rat, have three **eyelids**. The **upper** and **lower lids** are opaque and their outer surfaces are covered with hairy skin. The inner lid or **nictitating membrane** is transparent, with a slightly thickened rim. The skin lining the upper and lower lids, forming the nictitating membrane and covering the front of the eyeball, is very thin and transparent and is called the **conjunctiva**. It is kept in good condition by the secretions of the intraorbital and infraorbital lacrimal glands and the Harderian glands. That of the **lacrimal glands** is aqueous and excess is drained away into the nasal cavities through the lacrimal ducts. That of the **Harderian glands** is slightly oily and is especially important to swimming mammals.

EYEBALL

The eyeball is approximately spherical in most vertebrates, though it may be ovoid as in the rat, or have a distinct bulge on its outer surface as in man. Its wall is constructed of three layers of tissue.

(1) The **sclerotic** or outer layer is **fibrous** and **supports** and **protects** the soft inner layers. It is opaque at the back of the eye, where it is known as the **sclera**, but transparent in front, where it forms the **cornea**, which underlies the transparent conjunctiva, thus making a window into the cavity of the eyeball.

(2) The **choroid** or middle layer is highly **vascular** and, in most vertebrates, is pigmented, but the albino rat lacks pigment even here. The choroid is adherent to the sclera but separated from the cornea where it forms the **iris**. The iris has a central perforation, the **pupil**, and contains radial and circular bands of muscle by which the diameter of the pupil can be altered. In most vertebrates the iris thus acts as a shutter for reflex control of the amount of light entering the cavity of the eyeball. In the rat it is poorly developed, in keeping with the animal's nocturnal habits, and in the albino rat it is useless because lack of pigment renders it transparent.

Close to the periphery of the iris and to the corneo-sclerotic junction, the choroid layer is thickened to form the **ciliary body** with **ciliary processes** to which the lens of the eye is attached by the **suspensory ligaments**. In the rat the ciliary processes are usually double. The ciliary body contains the **ciliary muscles** used for focusing or **accommodation** of the eye.

(3) The **retina** or inner layer has a double structure. Its outer part is normally heavily **pigmented** to prevent internal reflection inside the eyeball. It extends over the ciliary processes to the limits of the iris, thus reinforcing the effect of the pigment in the choroid layer. In the albino rat it is, of course, unpigmented like the rest of the body.

The inner part of the retina extends around the back of the eyeball and ends at a line called the **ora serrata**. It consists of **light-sensitive cells, relay cells** and **ganglion cells** from which nerve fibres pass via the **optic nerves** to the brain.

The light-sensitive cells of vertebrate eyes are of two kinds—**rods** and **cones**. **Rods** contain visual purple or **rhodopsin**, a substance chemically allied to vitamin A. The quantitative, temporary bleaching of rhodopsin when exposed to light starts a chain reaction during which the nerve endings are stimulated. Rods have a low threshold of stimulation and thus respond in **dim light**. They react indiscriminately to light of different wavelengths so that the image is seen in **black and white**. **Cones** are of three types, each containing a special **opsin** activated by wavelengths in the blue, green and yellow regions of the spectrum respectively. Proportional stimulation of these enables the whole **colour** range to be appreciated. The threshold for stimulation of the opsins is higher than that for rhodopsin so that the cones can only function in **bright light**. In man the cones are most abundant in the region of the **fovea** at the centre of the back of the eye. The rat has no cones—an adaptation to its nocturnal habits.

There are neither rods nor cones where the optic nerve leaves the back of the eyeball. This is the **blind spot** and is always eccentrically placed.

CAVITY OF THE EYEBALL

The **lens** is suspended across the cavity of the eyeball and thus separates the front part from the back. The region in front of the lens is filled with watery fluid called the **aqueous humour** which allows free movement of the iris. The part in front of the iris is known as the **anterior chamber** and that between the iris and the lens as the **posterior chamber**. The region behind the lens is filled with gelatinous substance called the **vitreous humour** which serves to hold the retina and choroid in place against the sclera and to help to maintain the whole shape of the eyeball. The **lens** itself is a transparent or hyaline body composed of numerous flattened cells. It is biconvex in shape and has considerable elasticity.

The cornea, aqueous humour, lens and vitreous humour all contribute to the **focusing** of light on the retina to produce an **image** of the object viewed. Each has a different refractive index, therefore refraction takes place at each interface. The two sides of the cornea are parallel so that the effect of its refractive index is cancelled out, but its curvature affects the shape of the aqueous humour and therefore the focusing power of the eye. The curvature of the cornea is unalterable, as is that of the retina (except in some birds), but the shape of the lens can be changed to a different extent in different vertebrates. Such change, or **accommodation**, to view objects at different distances is reflex and is brought about by the **ciliary muscles**. When these muscles contract the tension of the **suspensory ligaments** is reduced and the lens swells under its own elasticity. This swelling reduces the radii of curvature of the surfaces of the lens and thus reduces the focal length so that a clear image of nearer objects is formed on the retina.

Man, like most of the arboreal mammals, has good powers of accommodation. His normal focal range is from infinity, when the eye is at rest, to about 250 mm distant, during maximum accommodation. In the rat the retinal distance is very short and the lens is almost spherical, with poor powers of accommodation, so that the animal is permanently short-sighted. This may be correlated with the fact that the animal's greatest activity takes place in dim lighting conditions when long-distance vision is impossible.

The focusing system of the vertebrate eye produces an **inverted image** on the retina, but the brain, on receiving impulses through the optic nerves, interprets the stimuli as a picture the right way up.

The degree of overlap of the fields of vision of the two eyes varies. Almost complete overlap, as in man, reduces the total area viewed but produces **stereoscopic** effects. Such stereoscopic vision is very important to all arboreal mammals which have to estimate the distances between the branches of trees. In the rat there is a certain amount of overlap and ability to use the eyes to judge distances, but a much greater part of the field of vision of each eye is independent, so that the total area viewed is very extensive.

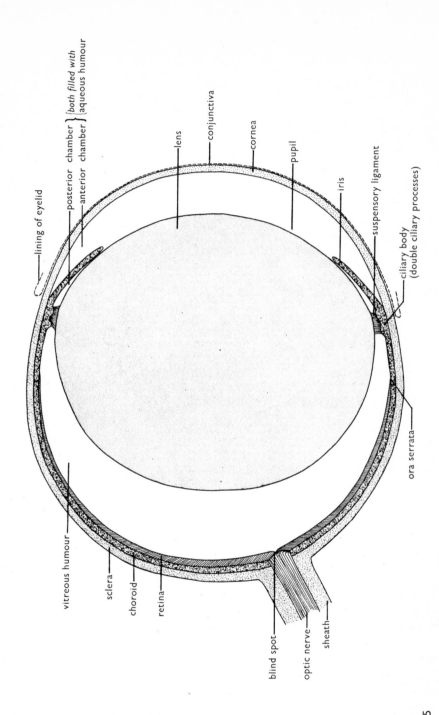

Fig. 94 **Eyeball (vertical section)**

lining of eyelid

posterior chamber ⎫ *both filled with*
anterior chamber ⎭ *aqueous humour*

lens

conjunctiva

cornea

pupil

iris

suspensory ligament

ciliary body
(double ciliary processes)

ora serrata

vitreous humour

sclera

choroid

retina

blind spot

optic nerve

sheath

× 15

HEARING AND BALANCE

The organs of hearing and balance are the **ears**. The parts of the ears concerned with balance are very similar in all vertebrates, but those concerned with hearing are more varied. In mammals they are very specialized and hearing is very acute. The mammalian ear has three distinct regions, the **outer ear**, the **middle ear** and the **inner ear**.

1. OUTER EAR

The outer ear consists of a tube called the **external auditory meatus,** from the surface of the head to the ear-drum or **tympanic membrane**. In the rat the outer part of this tube is dilated.

The external opening of the external auditory meatus is surrounded by a broad funnel, the **pinna**. The pinnae are supported by cartilage and vary in shape in different mammals. In most cases they can be moved by muscles so that the concave surface of each can be directed towards the source of sound. In man these muscles are vestigial and the pinnae cannot be moved appreciably.

The internal end of the external auditory meatus is covered by the delicate **tympanic membrane.**

2. MIDDLE EAR

The middle ear consists of a distended, air-filled **tympanic cavity** connected to the pharynx by a narrow passage, the **Eustachian tube**, through which pressure on the inside of the tympanic membrane can be equalized with that outside. The tympanic cavity is supported by the bony **tympanic bulla** and contains a chain of three small bones, the **malleus**, the **incus** and the **stapes** (see Auditory Ossicles, pages 4 and 10). The malleus is attached to the inner surface of the tympanic membrane and the stapes impinges on a small oval opening in the bone of the inner ear. This opening is the **fenestra ovalis** and close to it lies a round opening, the **fenestra rotunda**. Both openings are covered with membrane. The middle ear is generally stated to be homologous with the hyoidean (spiracular) gill slit of certain fishes, but recent work on its development and enclosure of the ossicles has thrown doubt on this.

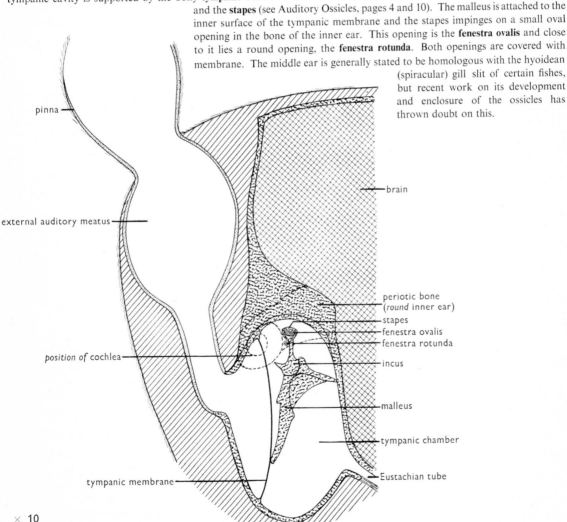

× 10

Fig. 95 **Ear (semi-sectional)**

78

3. INNER EAR

The inner ear consists of a **membranous labyrinth** surrounded by a **bony labyrinth** of similar shape formed by the **periotic bone**. The membranous labyrinth consists of **cochlea, saccule, semicircular canals, utricle** and **endolymphatic duct**. The bony labyrinth follows the cochlea, semicircular canals and endolymphatic duct closely, but is distended around the saccule and utricle to form the **vestibule**. The membranous labyrinth is filled with fluid called **endolymph**, while the spaces between the membrane and the bone are filled with similar fluid called **perilymph**.

The **cochlea** is spirally coiled. The **membranous cochlea** is connected with the small rounded **saccule** and is attached to the bone in such a way that three passages are formed, the **scala vestibuli**, the **scala media** and the **scala tympani**. The scala vestibuli and the scala tympani contain **perilymph**. They communicate with the **vestibule** and the region just inside the **fenestra rotunda** respectively. The scala media contains **endolymph**. It is separated from the scala vestibuli by the **vestibular membrane** and from the scala tympani by the **basilar membrane**. Both membranes are parts of the membranous cochlea. Inside the scala media is a stiff shelf-like fold called the **tectorial membrane** which overhangs the **organ of Corti**, a band of sensory and supporting cells on the basilar membrane. **Processes** projecting from the **sensory cells** are embedded in the tectorial membrane.

There are **three semicircular canals** in each ear. They are **mutually perpendicular**, i.e. they lie in different planes each of which is perpendicular to the other two. Each membranous semicircular canal lies close to one side of the corresponding bony canal and is attached at both ends to the **utricle**. At one end of each canal is a swelling or **ampulla**, the lining of which has patches of **sensory cells** with long projecting **processes**, called **cristae**. Patches of similar cells in the walls of the utricle and saccule form the **maculae**. In each macula the sensory processes are embedded in mucus which also holds numbers of small **calcareous granules**.

Nerve fibres from the sensory cells of the cochlea pass to a **spiral ganglion** and thence form the **cochlear nerve**. Fibres from the sensory cells of the ampullae and the utricle pass to small **vestibular ganglia** and thence form the **vestibular nerves**. The cochlear and vestibular nerves join to form the **auditory nerve** which goes to the brain.

Each membranous labyrinth develops from an **auditory vesicle** which is formed as an invagination on the side of the head of the embryo. The vesicle maintains its connection with the exterior for some time and the passage by which it does so becomes the **endolymphatic duct** when the aperture ultimately closes. The membranous labyrinth becomes surrounded by cartilage of the **auditory capsule** which is gradually replaced by the **periotic bone**.

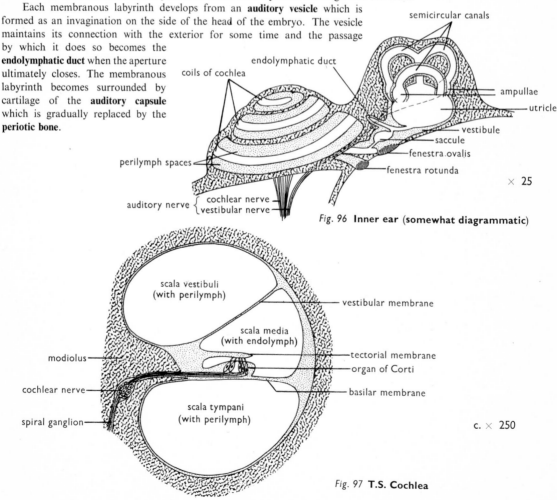

Fig. 96 **Inner ear (somewhat diagrammatic)**

× 25

Fig. 97 **T.S. Cochlea**

c. × 250

THE FUNCTIONING OF THE EAR

A. *HEARING*

Hearing is appreciation of the **vibrations** known as **sound waves**. **Pitch** of sound is governed by **wavelength** while **loudness** is governed by the **strength** of the waves. Sound waves can pass through any type of medium. Except in aquatic mammals, they reach the individual through the air. Sound waves in the air are feeble and have to be amplified before they can affect the fluids of the inner ear and thus produce stimulation of the sensory cells.

Sound waves reaching the **pinnae** are deflected into the **external auditory meati**, whence they pass to the **tympanic membranes**. The pinnae serve to concentrate the sound and thus improve acuity of hearing.

As each **tympanic membrane** vibrates it causes vibration of the **auditory ossicles**. These ossicles act as a system of **levers**, increasing the power of the vibrations, though reducing their amplitude, without affecting their wavelengths. By means of the auditory ossicles the vibrations are transmitted across the **tympanic cavity** to the **fenestra ovalis**, whence they affect the **perilymph**. Fluid is incompressible, but the membrane over the **fenestra rotunda** allows pressure in the inner ear to be equalized so that the delicate tissues are not damaged.

Vibrations in the perilymph are transmitted to the **endolymph** and result in vibrations of the **basilar membrane** of the **cochlea**. As this membrane vibrates the sensory cells of the **organ of Corti** are stimulated by tugging against the **tectorial membrane**. Different regions of the basilar membrane contain fibres of different lengths which resonate, or vibrate most actively, to sound waves of appropriate wavelengths. The localized effect on the organ of Corti results in discrimination of the **pitch** of sounds. The range of pitch appreciated varies in different types of mammals and also in different individuals. Thus some people can hear the squeaks of bats while others cannot do so. The auditory range of the rat is restricted to fairly high-pitched notes, in keeping with the high pitch of the voice.

Impulses from the organ of Corti are transmitted through the **cochlear nerves** and then the **auditory nerves** to the brain where they are interpreted.

Direction of the **source** of sound can be appreciated owing to the arrival of sound waves at the two ears independently. If the source is directly in front of or behind the head, the sound waves arrive at the two ears simultaneously, but if it is to one side or the other, they arrive at slightly different times. Though the individual does not notice two separate sets of stimuli, the brain records any difference and interprets it in terms of direction.

B. *BALANCE*

The individual is made conscious of position by two sets of stimuli, one of which is produced by movement, while the other is effective when the head is stationary.

When the head is **moved, fluid inertia** causes the endolymph in the **semicircular canals** to lag behind the movement of the walls of the canals. This lagging moves the processes of the **sensory cells** of the **ampullae**, producing stimulation of the nerve endings connected with them. Because the semicircular canals are **mutually perpendicular**, movement in **any plane** can be appreciated.

When the head is **still**, the **calcareous granules** of the **maculae** of the saccule and utricle are influenced by the direction of the pull of **gravity** and, according to the position of the head, rest on and stimulate the sensory cells with varying intensity. The macula of the saccule is affected when the head is tilted sideways and that of the utricle when it is tilted fore and aft. When the whole body is revolving rapidly, centrifugal force upsets this arrangement by acting on the granules in a direction different from the direction of gravity with resulting giddiness.

Both sets of stimuli result in impulses which are transmitted through the **vestibular nerves** and then the **auditory nerves** to the brain. They produce automatic or reflex adjustment of the tone of the muscles responsible for **posture**. The nerve centres concerned with balance are in the **cerebellum**.

EMBRYOLOGY

EMBRYOLOGY is the development of the individual from the time it is conceived, i.e. the time when the egg is fertilized. Many stages of the course of development are similar in all vertebrates but the details vary considerably even amongst nearly related species. The following account is generalized as much as possible to emphasize the peculiar features of mammalian embryology, but any details given apply to the development of the rat, which is highly specialized, particularly in the early stages.

If male and female rats are together mating or **copulation** will normally take place whenever the females are on heat. During mating **spermatozoa** are transferred to the **vagina** of the female by means of the **penis** of the male. They become motile and swim up the uteri to the **Fallopian tubes**. **Ovulation** takes place about eight hours after copulation and the eggs are swept into the Fallopian tubes by the cilia of the oviducal funnels. Each egg is surrounded by a **corona** formed by some of the cells of the follicle. This becomes partly mucified so that it does not impede penetration of the spermatozoa. **Fertilization** takes place within twenty hours of copulation and stimulates the completion of the second maturation division so that the second polar body is given off. Both the **egg nucleus** and the **sperm nucleus** then swell up and the two nuclei **fuse** together to form the **zygotic nucleus,** i.e. the first nucleus of the new individual.

Any eggs which are immature when ovulated, or are not fertilized, disintegrate within forty-eight hours. Those which are fertilized are retained within the Fallopian tubes for four days during which they develop into **blastocysts.** The corona is lost during the first twenty-four hours.

The first stage of development is **cleavage.** Each fertilized egg undergoes repeated cell divisions forming two cells (twenty-four hours), four cells (forty-eight hours), eight cells (seventy-two hours) and then sixteen cells (ninety-six hours). At about this time the sphincters at the tubo-uterine junctions relax and allow the young **blastocysts** to pass through into the uteri. Thereafter cell divisions proceed more irregularly. Each blastocyst becomes hollow and elongated with an **embryonal mass** of cells at one end.

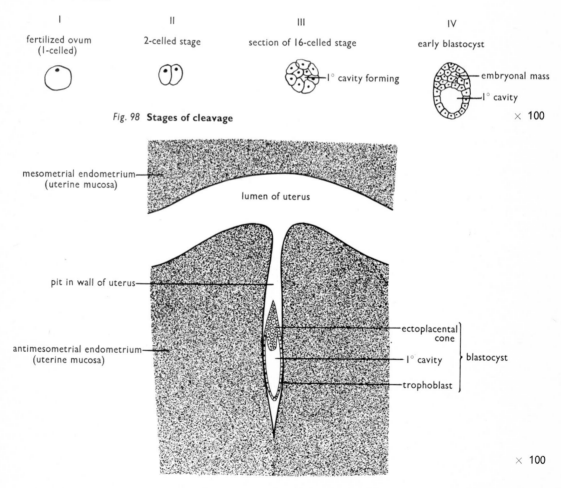

I
fertilized ovum
(1-celled)

II
2-celled stage

III
section of 16-celled stage

1° cavity forming

IV
early blastocyst

embryonal mass

1° cavity

Fig. 98 **Stages of cleavage**

× 100

mesometrial endometrium—
(uterine mucosa)

lumen of uterus

pit in wall of uterus—

antimesometrial endometrium—
(uterine mucosa)

ectoplacental
cone

1° cavity

trophoblast

} blastocyst

× 100

Fig. 99 **Newly implanted blastocyst** (*c. 6 days*)

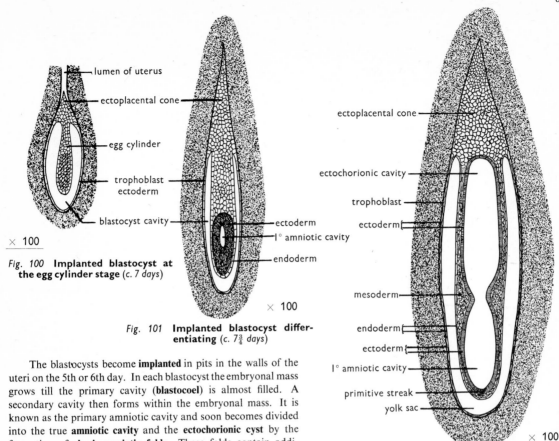

× 100

Fig. 100 Implanted blastocyst at the egg cylinder stage (c. 7 days)

Labels (Fig. 100): lumen of uterus; ectoplacental cone; egg cylinder; trophoblast ectoderm; blastocyst cavity

Fig. 101 Implanted blastocyst differentiating (c. 7¾ days)

× 100

Labels (Fig. 101): ectoderm; l° amniotic cavity; endoderm

Fig. 102 Implanted blastocyst at the primitive streak stage (c. 8½ days)

× 100

Labels (Fig. 102): ectoplacental cone; ectochorionic cavity; trophoblast; ectoderm; mesoderm; endoderm; ectoderm; l° amniotic cavity; primitive streak; yolk sac

The blastocysts become **implanted** in pits in the walls of the uteri on the 5th or 6th day. In each blastocyst the embryonal mass grows till the primary cavity (**blastocoel**) is almost filled. A secondary cavity then forms within the embryonal mass. It is known as the primary amniotic cavity and soon becomes divided into the true **amniotic cavity** and the **ectochorionic cyst** by the formation of **chorio-amniotic folds.** These folds contain additional cavities which fuse to form a cavity known as the **extraembryonic coelom** between the amnion and the ectochorionic cyst.

By this time the blastocyst is differentiated into **ectoderm, endoderm** and **mesoderm.**

The ectoderm forms:
(1) the **trophoblast** which covers the entire outer surface of the blastocyst and is thus adjacent to the uterine mucosa,
(2) the **ectoplacental cone,** a group of cells derived from the attached end of the embryonal mass,
(3) the lining of the ectochorionic cyst,
(4) the lining of the amniotic cavity.

The endoderm covers the embryonal mass and spreads to line the trophoblast so that the primary blastocoelic cavity becomes the **yolk sac.**

The mesoderm lies between the ectoderm and the endoderm in the region round the amniotic cavity and also lines the extraembryonic coelom.

The **embryo** develops from part only of the wall of the amniotic cavity while the rest of the blastocyst remains **extraembryonic** forming the **embryonic membranes.**

THE EMBRYO

The first sign of the embryo itself is the **primitive streak.** In *birds* and *reptiles* the primitive streak is the region where the mesoderm first becomes differentiated from the ectoderm and endoderm. In *mammals* the precocious development of the embryonic membranes is accompanied by much earlier differentiation of the mesoderm but the primitive streak is nevertheless formed as a temporary groove in the wall of the amniotic cavity. It marks the plane of bilateral symmetry of the embryo.

The anterior end of the primitive streak gives rise to a rod of cells which becomes the **notochord,** the primary skeletal structure. Anterior to this the ectoderm very soon becomes the **medullary plate** with two **medullary (neural) folds.** These extend posteriorly until the primitive streak is replaced. The folds approach one another and close over to form the **neural tube** from which the **brain** and **spinal cord** develop. In the rat the closure of the anterior end of the tube is delayed so that the head at first appears to be bilobed.

× 100

Fig. 103   **T.S. Primitive streak**

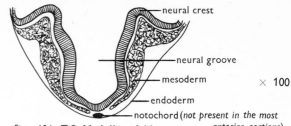

× 100

Fig. 104   **T.S. Medullary folds**

At this early stage the dorsal surface of the embryo is concave and the endoderm forms part of the wall of the yolk sac, so that there is no ventral surface. Closing in of the ectoderm to form the ventral surface of the embryo occurs by the formation of a **head fold** at the anterior end and a **tail fold** at the posterior end. The dorsal surface becomes gradually convex but the embryo retains connection with the extraembryonic parts at the **intestinal portal** or **umbilicus.**

Meanwhile the parts of the mesoderm near the neural tube become **segmented** and give rise to blocks of muscle while rudiments of all the principal organs are formed. The **heart** becomes functional at a very early stage and **blood** circulates in **blood vessels** in the embryo and in the embryonic membranes. The cartilaginous rudiments of the **skeleton** develop and then **ossify** gradually. Those of the vertebrae replace the notochord. Pronephric and mesonephric **kidney rudiments** are formed temporarily but are non-functional. **Metanephric kidneys** and also the **gonads** develop before birth. The **alimentary canal** and its associated **glands** become differentiated early but are non-functional till after birth. There is a stage during which the **pharynx** is perforated by **gill slits** and thus "*ontogeny recapitulates phylogeny*", i.e. development of the individual passes through a stage resembling an ancestral form. In this case the recapitulation is of the fish period of the evolutionary series. In the rat the **nervous system** becomes functional at about the end of the second week of pregnancy, though the **eyes** and **ears** are closed till after birth.

The details of all these developmental processes are beyond the scope of this work. The processes are governed by substances called **organizers** which resemble hormones in many ways and are functionally replaced by the **hormones** as soon as the endocrine organs are established. The development of the secondary sexual characters is a clear example of the action of hormones as organizers.

THE EMBRYONIC MEMBRANES

The **yolk sac, amnion, chorion** and **allantois** are **extraembryonic** structures which are known collectively as the **embryonic membranes**.

(1) YOLK SAC

The yolk sac is the structure which envelops the large mass of yolk in the developing eggs of *birds* and *reptiles*. It performs this function in *Monotremes* also but in other *mammals* it contains little or no yolk. When fully developed it consists of **ectoderm** and **endoderm** between which **mesoderm** penetrates. Groups of cells called **blood islands** form in the mesoderm and give rise to **blood vessels** at an early stage. When yolk is present it is digested by the endoderm of the yolk sac and is conveyed to the embryo in the blood in the **vitelline veins**. In non-yolky mammalian blastocysts the yolk sac circulation serves to transport materials absorbed from the wall of the uterus through the **trophoblastic ectoderm**. This arrangement forms the **yolk sac placenta** which is the only placenta of most *Marsupials*. In the majority of *Eutheria* the yolk sac placenta is relatively unimportant but in the rat the yolk sac mesoderm becomes invaded by maternal blood vessels and the whole structure is known as the **pseudo-placenta**.

(2) AMNION AND CHORION

In *birds, reptiles* and *Monotremes* and in many other *mammals* also, the amnion and the chorion develop from folds of the extraembryonic tissue round the embryo known as the **amniotic folds**. These folds close over in such a way that the inner layer, the **amnion**, ensheathes the embryo forming the **amniotic cavity** which is filled with **amniotic fluid** while the outer layer, the **chorion**, is continuous with the outer layers of the wall of the yolk sac. Both membranes consist of **ectoderm** and **mesoderm** only. The ectoderm of the amnion is internal, lining the amniotic cavity, while that of the chorion is external and, in mammals, is part of the **trophoblast**. The space between the amnion and the chorion is the **extraembryonic coelom**.

In the rat the development of these membranes is highly specialized and precocious, preceding the establishment of the embryo itself. Thus the amnion and the chorion are the transverse partitions which divide the cavity of the embryonic mass into three parts, see Fig. 106.

The **amnion** becomes distended to accommodate the growing embryo to which it remains attached at the **intestinal portal**. The **amniotic fluid** within it serves to protect the embryo against **mechanical shock**.

The **chorion** becomes associated with the **allantois** to form the **allanto-chorion**. In the rat it becomes fused to the **ecto-placental cone** so that the **ectochorionic cyst** is obliterated. This cyst has no homologue in forms with the simple, primitive method of formation of the chorion from the amniotic folds. When the chorion is established the outer wall of the blastocyst becomes known as the **chorionic vesicle**.

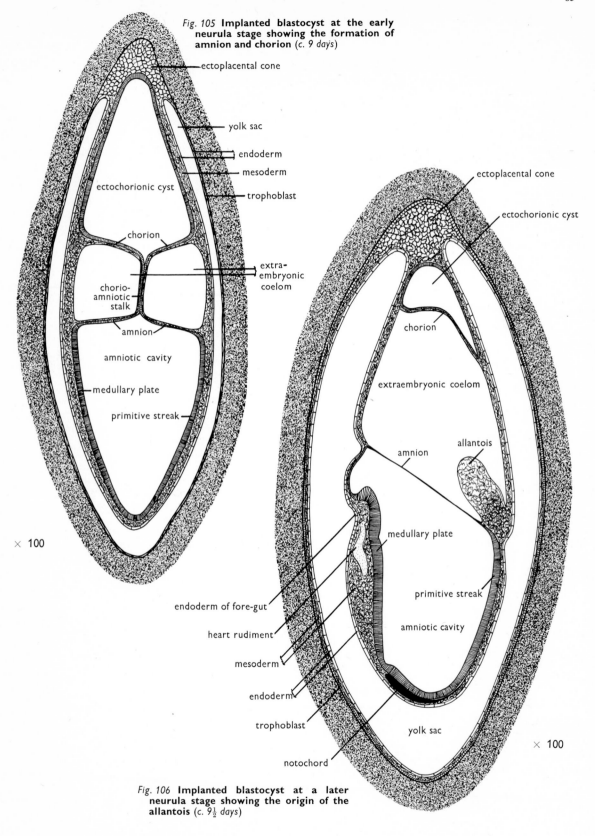

Fig. 105 **Implanted blastocyst at the early neurula stage showing the formation of amnion and chorion** (c. 9 days)

ectoplacental cone

yolk sac

endoderm

mesoderm

trophoblast

ectochorionic cyst

chorion

extra-embryonic coelom

chorio-amniotic stalk

amnion

amniotic cavity

medullary plate

primitive streak

× 100

ectoplacental cone

ectochorionic cyst

chorion

extraembryonic coelom

allantois

amnion

medullary plate

primitive streak

amniotic cavity

endoderm of fore-gut

heart rudiment

mesoderm

endoderm

trophoblast

yolk sac

notochord

× 100

Fig. 106 **Implanted blastocyst at a later neurula stage showing the origin of the allantois** (c. 9½ days)

(3) THE ALLANTOIS

The allantois of *birds* and *reptiles* is a diverticulum of the hind gut into which nitrogenous waste material in the form of uric acid is passed during the later stages of development within the egg shell. It consists of **endoderm** and **mesoderm** but no ectoderm, and grows out into the **extraembryonic coelom.** Part of its mesoderm combines with that of the chorion so that the **allanto-chorion** is formed. This is richly supplied with blood vessels and becomes closely applied to the egg shell where it serves as a respiratory surface.

In *mammals* the homologous structure is primitively hollow but in some cases, such as the rat, it is solid and has no endodermal core. It grows out from the posterior end of the embryo at a very early stage and extends across the extra-embryonic coelom to fuse with the chorion and form the **allanto-chorionic placenta.** This is the characteristic type of placenta of the *Eutheria*, though it is found in a few *Marsupials* also. It performs the functions of nutrition, respiration and excretion (see below).

THE PLACENTA

When the placenta is formed the embryo becomes known as the **foetus.** In the rat the **allanto-chorionic placenta** is not fully established till the end of the second week of pregnancy but the **pseudoplacenta** is functional some time before that.

The form of the true placenta varies in different mammals. In the rat it is **discoidal,** i.e. it is restricted to a disc-shaped area in the region originally occupied by the ectoplacental cone.

The placenta is formed from both maternal and foetal tissue. In the primitive condition the blood of the mother is separated from that of the foetus by the walls of the maternal blood vessels, uterine mucosa (connective tissue and epithelium), trophoblast (ectoderm of the chorionic vesicle), extraembryonic connective tissue, and the walls of the foetal blood vessels. In more specialized forms there is a considerable amount of erosion of these tissues though there is never any direct flow of maternal blood into the foetus, i.e. there is always at least one layer of tissue retained. In the rat there is a very intimate connection between the maternal and embryonic tissues and the only layer of cells which is retained in the completed placenta is the endothelium of the foetal blood vessels. This type of placenta is known as **endothelio-haemochorial.** In most other types with haemochorial placentae the trophoblast is also retained, e.g. in man.

When the foetal and maternal tissues grow into each other the superficial layer of the uterine mucosa becomes known as the **decidua.** In the rat the pits in which the blastocysts are implanted are **antimesometrial,** i.e. on the opposite side of the uterus to the mesovarium and the main uterine blood vessels. Implantation takes place with the ectoplacental cone directed towards the lumen of the uterus. As the blastocyst grows it fills the lumen of the uterus and the ectoplacental cone becomes connected to the **mesometrial** wall, so that the allanto-chorionic placenta is formed in the region of greatest blood supply.

The antimesometrial part of the decidua, i.e. that bordering on the yolk sac, accumulates lipoids and therefore the yolk sac circulation receives **fats** while the mesometrial part of the decidua, i.e. that connected to the true placenta, accumulates **glycogen** and therefore the allanto-chorionic circulation receives carbohydrates. It also receives **amino acids** from which proteins can be constructed and supplies of **oxygen** for the respiration of the foetal tissues.

When the kidneys start to function small amounts of **nitrogenous waste** materials are passed through the urethra into the amniotic cavity. The amniotic fluid never becomes more than equivalent to a very dilute urine, therefore much waste material must be removed by the placenta, through which there has been shown to be a very high rate of **water exchange.** **Carbon dioxide** is also removed through the placenta.

Thus the placenta serves for the **nourishment, respiration** and **excretion** of the foetus up to the time of birth. There is never any continuity of circulation between the mother and the foetus, so that the latter is protected against large molecules of potentially incompatible substances such as **proteins** and **hormones.** The placenta itself produces hormones—oestrogens, progesterone and chorionic gonadotropins—see page 62.

BIRTH

During the last part of the intra-uterine life the development is mainly concerned with growth. In the rat **organogeny** (establishment of the organs in potentially functional condition) is complete by the end of the second week. During the last week more blood is sent to the placentae of the smaller foetuses so that their growth rate is accelerated compared with that of the larger ones. Approximately 20% of the total birth weight is put on during the last 24 hours before **parturition.** The young are fairly uniform in size and state of development when they are born.

In the rat the young are born one at a time over a period of 1–2 hours. As each is delivered its embryonic membranes and associated decidua are shed also. The decidua and the foetal parts of the placenta are known as the **after-birth** and are immediately eaten by the mother. They serve to add some as yet unidentified substance to the milk without which the young do not thrive. The remaining investing membrane, the **amnion,** is then removed and the young rat starts to breathe. At the same time the two sides of the heart become sealed off from one another so that the double circulation is established, see page 45. Thus the change in circulation is coincident with the change in method of respiration.

The part of the **umbilical cord** which remains attached to the belly of the young soon shrivels up leaving only a small scar at the **umbilicus.**

The young rats are born naked, blind and unable to walk or fend for themselves. They are fed on milk till they are about three weeks old by which time they have grown hair, can see, hear, and run about, have teeth and can feed on a mixed diet similar to that of the adults.

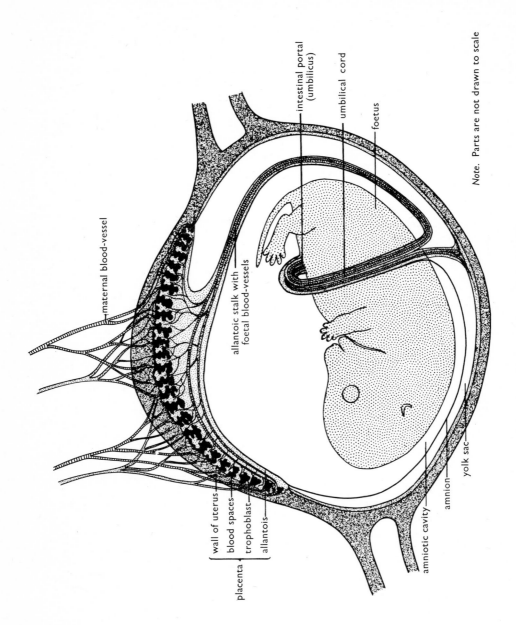

Note. Parts are not drawn to scale

Fig. 107 **Foetus with its membranes and placenta**

APPENDIX I

CARE AND BREEDING OF RATS

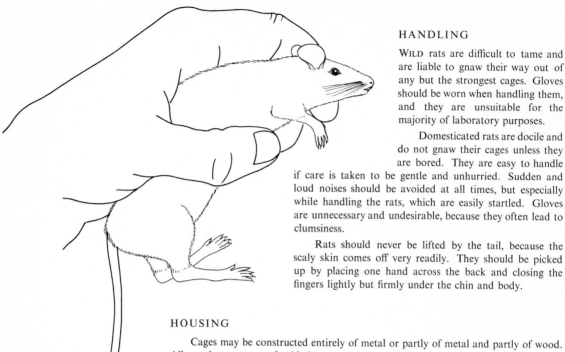

HANDLING

WILD rats are difficult to tame and are liable to gnaw their way out of any but the strongest cages. Gloves should be worn when handling them, and they are unsuitable for the majority of laboratory purposes.

Domesticated rats are docile and do not gnaw their cages unless they are bored. They are easy to handle if care is taken to be gentle and unhurried. Sudden and loud noises should be avoided at all times, but especially while handling the rats, which are easily startled. Gloves are unnecessary and undesirable, because they often lead to clumsiness.

Rats should never be lifted by the tail, because the scaly skin comes off very readily. They should be picked up by placing one hand across the back and closing the fingers lightly but firmly under the chin and body.

HOUSING

Cages may be constructed entirely of metal or partly of metal and partly of wood. All-metal cages are preferable because they are easier to clean and disinfect. The most convenient type of metal cage is made of galvanized iron wire and has a detachable wire floor over a solid metal débris tray. This allows cleaning to be done without disturbance of the occupants of the cage and reduces the chances of fouling the food, thus reducing the spread of intestinal parasites. The débris pan should be deep enough to prevent the rat from reaching its contents through the floor, and should contain softwood sawdust, peat or any other suitable absorbent material. Care should be taken that any material used has not been contaminated by vermin, cats or dogs. It should be sterilized immediately before use if there is any chance of contamination during transport or storage. Whenever possible the débris should be cleared away daily, and it should be burnt immediately because it attracts flies.

If the cage has a wire floor there should invariably be a small resting platform. A shelter, such as an inverted tin with a hole in one side, should be placed over the platform, and suitable bedding such as shredded paper or sterilized hay should be provided. The rat rarely fouls its sleeping quarters, but they should be cleaned out once a week and the used bedding burnt. At the same time the whole cage should be washed down with soapy water, and it should be disinfected at least twice a year.

Breeding cages usually have solid sides and top to reduce draughts. Unless laboratory conditions are optimal, they should be provided with wooden nesting boxes approximately 200 mm square. Each such box should have a small opening in one side only and ample bedding material. Breeding cages should whenever possible be left undisturbed till three weeks after the birth of the young, but must be thoroughly cleaned and disinfected before they are used again.

Rats are active animals and unless they have sufficient exercise they become bored and unhappy. Cages should therefore be as large as convenience allows. The minimum cage size for normal housing of groups of rats is $750 \times 450 \times 450$ mm. Mating and breeding cages are usually smaller.

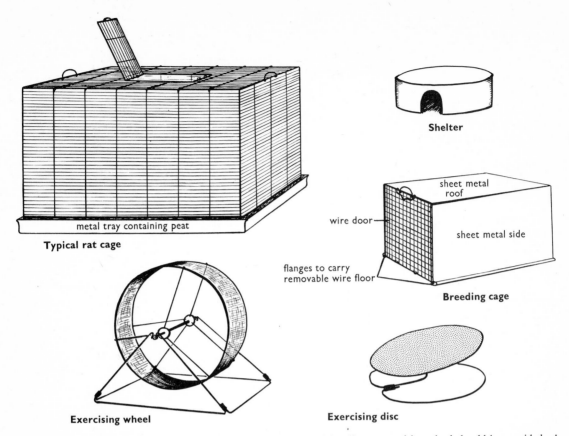

Typical rat cage

metal tray containing peat

Shelter

sheet metal roof

wire door

sheet metal side

flanges to carry removable wire floor

Breeding cage

Exercising wheel

Exercising disc

If the cage has to be small owing to shortage of space, an exercising disc or exercising wheel should be provided. An exercising disc may be made of plywood, pressed fibre or plastic material. It pivots on a stand, which should be larger than the disc itself in order to prevent the disc from catching against the sides of the cage. An exercising wheel should have a diameter equal to at least three times the length of the head and body of the rat and clearance under the wheel of at least the diameter of the body. It is usually made of strong wire gauze on a wire or wooden frame. All such exercising devices require cleaning and periodic disinfection, like the cages.

Except while pregnant, rats should not be kept alone. They thrive better in pairs, even when of the same sex. It has also been shown that even numbers do better than odd numbers when several rats are caged together. They are not normally quarrelsome animals, but it is better to let the males establish themselves before introducing females into the same cage

ROOM CONDITIONS

(i) Temperature

Ordinary laboratory conditions are usually suitable for the keeping of rats, provided there are no draughts and no sudden changes in temperature. Slight variations are valuable for stimulating vigour but extremes should be avoided. Excessive heat is more dangerous than cold and lowers resistance to disease. Nevertheless special animal rooms are usually kept warmer than the ordinary laboratory, i.e. between 18·5 °C and 24 °C, and extra warmth (24·5–25·5 °C) is advisable for breeding.

(ii) Light

If cages are near windows they must be screened from direct sunlight during hot weather. A sheet of cardboard between the window and the cage or over the top of the cage is sufficient. The winter decrease in fertility is probably a response to seasonal reduction in the amount of light. Artificial light can be used as a substitute to promote regular breeding throughout the year. Light also affects the time taken to reach sexual maturity. Provided a shelter is available the cage should be as light as possible.

(iii) Humidity

Good ventilation is essential, because stagnant air increases the tendency to develop pneumonia. A relative humidity of about 50 is best. Dry air causes ringtail in young rats, while excessive dampness shortens the life of the adult.

FEEDING

Rats may be given a wide variety of food.

Prepared food pellets are available and may be obtained from

Oxoid Ltd
20 Southwark Bridge Road
London, SE1 9HF

Harris Biological Supplies Ltd
Oldmixon Crescent
Weston-super-Mare, BS24 9BJ

Gerrard and Haig Ltd
Gerrard House, Worthing Road
East Preston, Sussex, BN1 6AS

Charles River (U.K.) Ltd,
Manston Road,
Margate, Kent, CT9 4LT

Food hopper

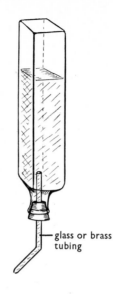

glass or brass tubing

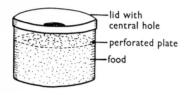

lid with central hole
perforated plate
food

Non-scatter feeding dish

Water bottle

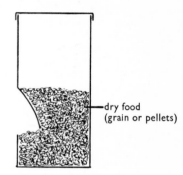

dry food (grain or pellets)

Food hopper—sectioned

If stored in a tin away from damp, such pellets will remain fresh for a considerable period, but they should not be kept indefinitely. They are simple to use, economical and not readily fouled. The rats like holding them in their "hands" and should be allowed to make stores which can be removed from time to time when the cage is cleaned. Enough pellets to last over the week-end may be left in a hopper such as that shown in diagram.

Pellet food is somewhat monotonous and should be supplemented by other things such as carrots, green vegetables, meat, liver, kidneys, heart, fish and skimmed milk. Rats do not like onions or potato peelings, though they will eat whole potatoes readily enough.

If pellets are not used, the diet should include a considerable amount of carbohydrate in such form as wheat, oat and barley flour and ground maize, with optional soya flour, lentils or peas. (*Note.* Coarse-ground maize may cause ulcerated stomach in newly weaned rats.) Fat should be supplied by linseed meal, and any protein additional to that provided by the pulse food should be supplied by skimmed milk powder, fish meal and occasional meat supplement. The food should be adequately salted and a few drops of cod-liver oil should be given twice weekly to supply vitamins A and D.

A basic diet consisting of the following has been considered adequate by a number of authorities:

10–15% linseed meal
10–15% dried skimmed milk
60–70% wheat and oat flour, with sometimes barley flour and ground maize
up to 20% lentils, peas or soya flour
1% salt

A certain amount of green food or root vegetables is usually added, and canned or fresh meat given twice a week. Fresh milk is also recommended for nursing does and for young rats, but no milk should be given to the doe from the time of weaning of her young until she has stopped lactating.

Food and milk should be given in vessels which are not easily upset, e.g. the heavy jars used for face creams. Non-scatter feeders are available.

Plenty of fresh water should be given in containers in which it cannot be contaminated. The best arrangement is a bottle outside the cage with a long brass or glass mouthpiece projecting inwards about 140 mm above the floor for adults and 70 mm for young. The mouthpiece has a small hole so that the water can be obtained by sucking.

Rats in captivity do not have to gnaw to obtain their food, and therefore there is a tendency for the incisor teeth to become overgrown. A wooden block should be supplied, but if gnawing this is still insufficient to wear the teeth down as rapidly as they grow, bone forceps may be used to clip them.

BREEDING

Rats are very easy to breed. A number of males and females may be run together and the latter removed when they show signs of pregnancy, such as marked increase in weight, enlargement of the abdomen and a droplet of blood at the vaginal opening.

The females come on heat every 4 or 5 days. The fertility decreases between October and December and then rises again to normal by the end of February. It is, however, possible to get litters all the year round.

Signs that the female is coming on heat and is ready for mating are:

(i) an increased running activity with nervous darting about;
(ii) ear quivering when the head is stroked;
(iii) bracing of the body when the perineal region is touched;
(iv) change in the appearance of the superficial genitalia which become dry, ridged and bluish in colour instead of moist, smooth and pink.

The activity signs usually show in the evening because the rat is by nature nocturnal.

Though the rat becomes sexually mature when between 50 and 60 days old, it is best to delay the first mating till it is 100 to 120 days old. It is also best to mate animals of approximately the same age. In many laboratories brother-and-sister matings are practised regularly. This helps to maintain the purity of the stock without any apparent ill effects.

The gestation period is 21–23 days, enlargement of the abdomen becoming obvious on about the 14th day. The gestation period may be extended up to a week longer than the normal period if the rat is carrying a large number of young and feeding the previous litter at the same time, but this condition should not be permitted except in special circumstances. The female should be isolated until she has finished nursing and should be rested for at least 2 weeks before the next mating. Occasionally double littering occurs without the renewed presence of the male, a fortnight intervening between the litters.

Parturition usually takes 1–2 hours, but the time varies with age, size of litter and the physical condition of the mother. Signs that it is about to begin are stretching of the body and hind legs so that the latter are lifted off the floor of the cage and change in form as the young descend to the lower part of the abdomen. Normally, the rat licks the vulva immediately before delivery and absence of this sign often indicates that something is wrong. During delivery she stands half crouched while the foetus slips out still covered in its amniotic sheath. She then pulls out the placenta and eats it before freeing the young and licking it. It has been shown that the placenta contributes a growth-promoting substance to the milk.

The first member of the litter usually emerges head first, but the others may come either way. After they have all been born, the mother assembles them in the nest, cleans herself and permits nursing. Throughout the process she should not be disturbed. She must never be taken from her young while she is actually nursing them or she may bite and will probably eat the young. Maternal care is also affected by noise.

The number of young varies, increasing up to the fifth litter and then decreasing. The general average is 8 and litters of more than 10 should be reduced by removal and euthanasia of those which appear weakest. When many rats are being bred, it is often possible to put some of the young from a large family to a foster-mother whose own family is small. When this is done the foster-mother must be excluded from her nesting box until the alien young have had time to acquire the nest odour. Otherwise they are always killed.

The new-born rat is hairless and blind, with closed ears, undeveloped limbs and a very short tail. It makes wriggling and paddling movements but cannot crawl. Its head is continually searching and shows quick response to smell and taste. The sexes can be differentiated because the male has a larger genital papilla and a greater distance-between this papilla and the anus.

The ears open on the 2nd or 3rd day and the incisors erupt between the 8th and 10th days. By about the 10th day the young can crawl supported by their limbs, and their eyes open between the 14th and 17th days. The nipples of the female develop between the 8th and 15th days, and they, and the other genitalia, are obscured by hair by the 16th day. The molars erupt on the 19th, 21st and 35th days respectively. Thus almost all the teeth show before the time of weaning, which is usually done by removing the young from the mother and putting them in a separate cage on the 21st day. If permitted to do so, the mother will feed the young for several days longer.

The diet of young rats is much the same as that of older rats, but with supplementary milk.

The life-span of the albino rat is about 3 years, but is reduced by high humidity and extremes of temperature. In spite of this relatively long life, which is reckoned to be equivalent to 90 years in a human being, a male rat should not be used for breeding after it is 14 months old and a female should not be bred from after she is about 12 months old. She should not have more than 7 litters altogether, and she undergoes the menopause when she is 15 to 18 months old.

PARASITES

Rats are subject to both external and internal animal parasites, and care should be taken to prevent infestation as far as possible and cure it if it does occur.

External parasites include fleas, lice, bed bugs and mites.

Fleas and **lice** occur on the rat itself and may be treated with pyrethrum dust, D.D.T. or (except for young rats) derris and rotenone. Painting with larkspur solution will kill lice and their nits, and the latter need not then be removed from the hair. The cages should be washed with lysol or paraffin oil emulsion. This will kill the eggs, larvae and pupae of the fleas.

Bed bugs only come on the rat to feed and can usually be removed by thorough cleaning and painting the cages with insecticide, but fumigation may be necessary if the infestation is severe.

Mange mites burrow into the skin and can be treated with sulphur ointment or benzyl benzoate emulsion.

Blood-sucking mites may be treated with larkspur solution or light anaesthesia of the rat, which will cause the mites to come to the surface of the fur, from which they can then be brushed off. The bedding should also be changed.

Flies and **cockroaches** are not parasites, but they may be attracted by the food and débris of rats. Many poisons which are used for flies are also poisonous to the rats, so that care must be taken not to contaminate the food. The best preparation to use is D.D.T. as a powder or as a paint with paraffin.

Internal parasites include tapeworms, Nematodes and numerous types of Protozoa.

Tapeworms are usually highly specific regarding their hosts. The commonest species found in rats can have both adult and cysticercus stages in the same host. The eggs hatch in the intestine. The onchospheres invade the tissues and become cysticerci, but leave the tissues after one to two weeks and become mature when they reach the intestine again. The adults do not live long (max. 7 weeks), but reinfection is easy unless faecal contamination of the food is prevented.

Although an intermediate host is not essential, a cysticercoid stage may develop in fleas and in mealworms, which are thus capable of spreading the parasites and are frequently the source of the original infestation of the rat colony.

One of the tapeworms found in cats has a cysticercus stage in the liver of the rat which may cause tumours.

Nematodes have very varied life-histories and the numerous species which occur in rats enter the body in a number of different ways, e.g. *Strongyloides* through the skin, *Trichinella* by cannibalism, *Trichosomoides* by food and water contaminated by urine, and *Gongylonema* by eating the insects which are the intermediate hosts. Many of these Nematodes have their adult stages in the intestine of the rat and their eggs are present in the faeces, but some attack other parts, e.g. *Trichosomoides* in the bladder and *Trichinella* in the muscles. Infestations (especially of Trichinella) may, if severe, cause disease and death, but if the rat recovers it is usually immune to further attacks of the same parasite.

Protozoa of many different kinds are found in the rat. Most of those which inhabit the gut are non-pathogenic and live on the gut contents without causing disease. A few cause ulceration, e.g. *Entamoeba histolytica*, which also causes amoebic dysentery in man.

Blood parasites of rats include Trypanosomes, e.g. *Trypanosoma lewisi*, which is transmitted by the rat flea.

It is obvious from the above description of the internal parasites of rats that the commonest causes of infection are food and water contaminated by faeces and urine, bedding which has been contaminated by other animals, and infestation by external parasites which may in turn come from contaminated bedding. Strict hygiene is therefore important in reducing the incidence of diseases produced by these parasites.

DISEASES

Besides the diseased conditions produced by the parasites already described, the rat is subject to a number of other infections.

Paratyphoid*of rats is caused by bacteria of the*Salmonella* group. The acute form of the disease is usually fatal, but there is also a chronic form without obvious external symptoms. Individuals which have this act as carriers from whom the disease may be spread by contamination of the food. The organisms which cause this disease also cause food poisoning in man.

Middle ear disease is very common, but symptoms are only seen when inflammation spreads to the labyrinth of the inner ear and causes characteristic difficulty in balancing and lack of co-ordination of the muscles. Although various bacteria have been isolated from discharges of the middle ear, it has not been proved that any of them are responsible for the disease. Rats under three months old are usually immune and strains which are free from the disease have been developed.

Pneumonia is the commonest disease of adult rats. The symptoms are an unhealthy appearance and rapid, noisy breathing. There is no evidence that specific bacteria are concerned and the predisposing causes of the disease are unsatisfactory living conditions, including:

    (i) stagnant air, i.e. insufficient ventilation;
    (ii) dusty air;
    (iii) accumulation of débris;
    (iv) stale bedding.

All animals known to be suffering from the above diseases should be destroyed and their cages disinfected immediately.

Rats are also subject to non-infectious tumours, both malignant and benign, and to diseases of the heart, blood-vessels and kidneys, including arteriosclerosis. Such diseases are often responsible for the ultimate death of the animal in old age.

* Human paratyphoid is due to a completely different species of bacterium, *Bacillus paratyphosus*.

ANAESTHESIA

Anaesthesia is necessary for all experiments likely to cause pain to the animal, especially all operative procedures. Such experiments may be performed only by persons holding a **Home Office Licence** and do not come within the scope of ordinary student work. The following information is therefore solely for interest and must not be used by any unlicensed individual.

Chloroform is unsuitable for anaesthetization of rats. **Ether, avertin, nembutal** and **urethane** are all used according to the depth of anaesthesia required and whether the rat is to be allowed to recover or not.

Ether gives shallow anaesthesia, with recovery in 5–10 minutes. The ether is applied on cotton wool, which is placed in a beaker over the head of the animal. The ether-soaked wool must not touch the nose of the rat because it may cause fatal shock. Violent convulsions frequently occur during the early stages of anaesthetization, and occasionally there is primary shock requiring artificial respiration. When the rat is anaesthetized the ether should be removed, but it should be reapplied at once if there are signs of recovery before the operation is complete.

Avertin (tribromethanol) is prepared as a 2·5% solution in saline at 60°C. The dose injected is graded according to the size of the rat, thus:

30– 60 g rat	0·6 mg per 100 g
60–100 g rat	0·7 mg per 100 g
over 100 g rat	0·8 mg per 100 g

Nembutal in saline solution is similarly injected, the dosage being 2·85 mg per 100 g body weight. It produces much longer-lasting anaesthesia (2–3 hours), but the rat takes about 20 minutes to become fully anaesthetized.

Urethane is used only when there is to be no recovery. Recently the genuineness or completeness of the apparent anaesthesia has been questioned. There is a possibility that the animal may be able to feel pain though showing no reaction. Pending further investigation this drug should not be used unless there is absolutely no alternative: e.g. for a very lengthy operation. The dosage is 0·6 ml of a 25% solution per 100 g body weight. The induction is even slower than for nembutal.

EUTHANASIA

Euthanasia means, strictly, putting to death painlessly; but with respect to human beings the term is commonly used to mean mercy killing of the incurably sick, etc.

Humane killing of rats is best performed by **ether** or **coal gas**. The rat is placed in a dry tin with a perforated lid. It is important to permit air to reach the animal so that it is not asphyxiated. Ether is poured on a cotton wool pad which should be suspended inside the tin so that it does not touch the rat, or, alternatively, a tube from the gas supply is led into the tin and the gas is turned on slowly. When the rat becomes unconscious the air-holes may be covered over to hasten death. The rat should not be removed till rigor mortis has set in.

Note. Both ether and coal gas are extremely **inflammable**, so that there should be no naked flame or lighted cigarette anywhere near the killing chamber.

If the animal is not required for dissection it may be killed by a single heavy blow on the head. There must be no hesitation, because a glancing or feeble blow may cause pain. Until the operator has had sufficient experience to be certain of killing the rat at the first blow, it is advisable to give preliminary anaesthetization with ether. The best method of applying the blow is to hold the rat by the root of its tail and swing it very hard against a table or sink.

APPENDIX II

GENETICS

GENETICS is the study of the mechanism of heredity.

The early work on this subject is known as **Mendelism** because the first controlled experiments were those of the Abbé Mendel in the latter half of the nineteenth century. Mendel used the garden pea and obtained results from which he postulated two laws. Subsequent work has shown that these laws are applicable to animals as well as to plants and has extended our understanding of the mechanism of heredity far beyond Mendel's theories.

Mendel believed that the development of distinctive characteristics was due to the presence in the individual of definite **factors** for these characteristics. In his **first law** he stated that the **gametes** are the bearers of these factors from one generation to the next and that any alternative factors affecting a chosen character are mutually exclusive. Thus a gamete can carry only one member of a pair of factors while the zygote formed from fusion of two gametes bears two. Two like pure-bred parents will give like factors to their offspring so that these offspring contain two identical factors and are themselves **pure bred** for the character concerned. Two unlike pure-bred parents will give unlike factors to their offspring so that these offspring contain two factors which are not identical and are **hybrid** for the character concerned. In any hybrid individual the factor which produces the visible effect is said to be **dominant** while the other is **recessive**. Occasionally neither is dominant and an **intermediate** appearance is produced.

Mendel's factors are now known as **genes**. A pair of alternative factors are **allelomorphic genes** or **alleles**. The genetic constitution of an individual is its **genotype**, while the appearance of the individual is its **phenotype**.

Mendel's first law may be illustrated by consideration of the inheritance of *fully coloured* coat versus white or *albino* coat in rats.

If a pure-bred coloured rat is crossed with a pure-bred albino rat the resultant offspring are all coloured, indicating that colour is dominant and albino is recessive. This occurs irrespective of the sex chosen to be the coloured parent.

If the symbol $[C]$ stands for the gene for coloured coat and the symbol $[c]$ stands for the gene for albino coat, then according to Mendel's law the pure-bred coloured rat must have the genotype $\frac{C}{C}$ because it received the factor or gene for colour from each of its coloured parents, while the pure-bred albino rat must similarly have the genotype $\frac{c}{c}$.

The coloured rat produces gametes bearing $[C]$ and the albino rat produces gametes bearing $[c]$.

Thus the hybrid rats resulting from the cross must have the genotype $\frac{C}{c}$, though they are of the coloured phenotype.

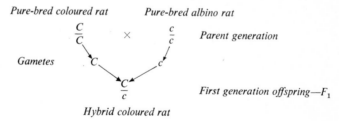

Pure-bred coloured rat   *Pure-bred albino rat*

$$\frac{C}{C} \quad \times \quad \frac{c}{c} \qquad \textit{Parent generation}$$

Gametes C c

$$\frac{C}{c}$$

$\textit{First generation offspring—F}_1$

Hybrid coloured rat

If the hybrid coloured rats are now inbred, there are coloured and albino rats among the offspring and the ratio of the two types is approximately 3:1. This result can also be explained by the assumptions made in Mendel's first law.

Thus the hybrid rats produce gametes which may bear either $[C]$ or $[c]$, but not both together. The two types of gametes are produced in equal numbers and unite in pairs indiscriminately. Thus:

Gametes	C	c
C	$\frac{C}{C}$	$\frac{C}{c}$
c	$\frac{C}{c}$	$\frac{c}{c}$

i.e. $1\frac{C}{C} : 2\frac{C}{c} : 1\frac{c}{c}$

or   3 coloured : 1 white

It is obvious that the coloured rats are of two types. The simplest method of demonstrating this difference is the **test cross**, in which each coloured rat is mated to an albino rat. All the offspring of the pure coloured rats $\frac{C}{C}$ must be coloured as in the original cross, while half the offspring of the hybrid rats $\frac{C}{c}$ are coloured and the other half are albino.

Thus:

Similar results can be obtained for a number of pairs of characters affecting rats, the symbols for the genes for which have been standardized thus:

Agouti (wild grey colour) [A] is dominant to non-agouti (any other colour) [a]
Black [B] is dominant to chocolate [b]
Curly coat [Cu] is dominant to normal coat [cu]
Intense colour [D] is dominant to dilute colour [d]
Self-coloured coat [H] is dominant to hooded [h]
Normal coat [Hr] is dominant to hairless [hr]
Normal coat [K] is dominant to kinky coat [k]
Fully coloured [P] is dominant to pink-eyed, yellow [p]
Fully coloured [R] is dominant to red-eyed, yellow [r]
Normal gait [W] is dominant to waltzing gait [w]
Normal [L] is dominant to lethal [l]

An individual which is pure bred for a given character is said to be **homozygous** for the genes concerned, while one which is hybrid is said to be **heterozygous**. Any dominant gene shows its effect whether the individual is homozygous or heterozygous, but any recessive gene can only show when it is homozygous.

In the above list the recessive lethal [l] can only take effect in the homozygote $\frac{l}{l}$ and such an individual will in fact never be born. The recessive hairless [hr] is not lethal, but in the homozygous condition $\frac{hr}{hr}$ it has some deleterious effect on survival other than preventing growth of hair. On test-crossing hybrids $\frac{Hr}{hr}$ fewer of the young show this character (3 : 5) than would be expected in the case of an ordinary recessive (1 : 1). Death of the weakened young occurs soon after birth, but before hair has started to grow, so that it is not a direct effect of the hairlessness.

Mendel's **second law** or the **law of independent assortment** states that pairs of characters are inherited entirely independently of one another. More recent work has shown that this is true in certain cases only, while in others there is **linkage**.

As an example of **independent** assortment in the rat the two pairs of characters agouti/non-agouti and curly/smooth may be considered together.

Thus:

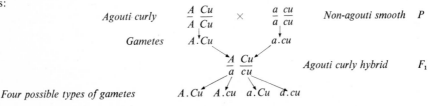

Chance inbred mating of the hybrid would result in the following gene combinations:

	A.Cu	A.cu	a.Cu	a.cu
A.Cu	$\frac{A\ Cu}{A\ Cu}$	$\frac{A\ cu}{A\ Cu}$	$\frac{a\ Cu}{A\ Cu}$	$\frac{a\ cu}{A\ Cu}$
A.cu	$\frac{A\ Cu}{A\ cu}$	$\frac{A\ cu}{A\ cu}$	$\frac{a\ Cu}{A\ cu}$	$\frac{a\ cu}{A\ cu}$
a.Cu	$\frac{A\ Cu}{a\ Cu}$	$\frac{A\ cu}{a\ Cu}$	$\frac{a\ Cu}{a\ Cu}$	$\frac{a\ cu}{a\ Cu}$
a.cu	$\frac{A\ Cu}{a\ cu}$	$\frac{A\ cu}{a\ cu}$	$\frac{a\ Cu}{a\ cu}$	$\frac{a\ cu}{a\ cu}$

Thus, of every 16 offspring:

9 are agouti curly (i.e. show both the dominant characters)
3 are agouti smooth (i.e. show one dominant and one recessive character)
3 are non-agouti curly (i.e. show the other dominant and the other recessive character)
1 is non-agouti smooth (i.e. shows both recessive characters).

Notice that summated there are 12 agouti: 4 non-agouti, i.e. 3:1, and that there are 12 curly: 4 smooth, i.e. also 3:1.

Notice also that only one individual of each of the four phenotypes is homozygous for both pairs of characters, while the others are heterozygous for one or other or both pairs of characters.

As an alternative to inbreeding, a test cross can be made using a non-agouti smooth-haired rat to mate with the F_1 hybrids. In this case the four possible types of offspring are produced in equal numbers.

	$A.Cu$	$A.cu$	$a.Cu$	$a.cu$
$a.cu$	$\dfrac{A\ Cu}{a\ cu}$	$\dfrac{A\ cu}{a\ cu}$	$\dfrac{a\ Cu}{a\ cu}$	$\dfrac{a\ cu}{a\ cu}$

Note. Throughout the above experiments the same result would have been obtained if the original cross had been between an agouti smooth-haired rat and a non-agouti curly-haired rat, because there is no linkage.

In the above example of independent assortment the genes concerned affect different types of characteristic, i.e. colour and texture of the hairs. Sometimes two or more pairs of genes affect the same characteristic and therefore interact with one another. As examples of **interaction** of **factors** affecting coat colour of rats, any individual will be albino unless it carries the colour gene [C] which allows the development of any colour at all, and it will be agouti if it carries the agouti gene [A] irrespective of any other colour genes it may contain. As a result of this the appearance of the offspring may be very different from what would be expected if simple independent non-interacting factors were concerned.

EXAMPLE 1. An albino rat carrying the agouti genes $\dfrac{A\ c}{A\ c}$ is crossed with a non-agouti coloured rat $\dfrac{a\ C}{a\ C}$. The colour of the latter is immaterial provided that both rats are homozygous for the same factor, e.g. [B] throughout would make the coloured rat black while [b] throughout would make it chocolate-coloured.

Then:

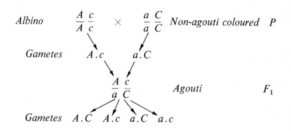

On inbreeding such agouti rats the offspring are:

9 agouti : 3 non-agouti coloured : 4 albino.

Notice that though three of the albino rats carry the agouti gene [A] they are unable to show it in the absence of the colour gene [C].

On test-crossing the F_1 agouti rats, the offspring are, for the same reason:

1 agouti : 1 non-agouti coloured : 2 albino.

EXAMPLE 2. Pure-breeding black and agouti rats must both be homozygous for the colour gene C, otherwise they would have occasional albino offspring. On crossing such rats, therefore, the colour gene can be ignored. The black rat must be homozygous for the non-agouti gene in order to show the black, and if it is true breeding it must also be homozygous for the black gene. Thus it has the genetic constitution $\dfrac{a\ B}{a\ B}$. The agouti rat must be homozygous for the agouti gene in order to breed true for the grey coloration, but it may or may not be carrying the black gene masked by the agouti one. Thus it may have the constitution:

(i) $\dfrac{A\ B}{A\ B}$ or (ii) $\dfrac{A\ b}{A\ b}$ or (iii) $\dfrac{A\ B}{A\ b}$.

(i) If the agouti rat is $\dfrac{A\ B}{A\ B}$ then the cross with the black rat $\dfrac{a\ B}{a\ B}$ is in effect a monohybrid cross between agouti and non-agouti individuals, i.e. it concerns one pair of characters only, and the F_2 from inbreeding produces 3 agouti : 1 black. The gene for blackness could in fact be ignored throughout, as postulated in example 1 above.

(ii) If the agouti rat is $\dfrac{A}{A}\dfrac{b}{b}$, then the cross with the black rat $\dfrac{a}{a}\dfrac{B}{B}$ is an ordinary dihybrid cross, i.e. it concerns two pairs of characters both of which are homozygous in the parents.

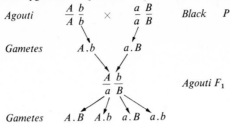

The F_2 produced by inbreeding of the F_1 agouti rats consists of:

12 agouti : 3 black : 1 chocolated-coloured.

The 12 agouti rats represent $9+3$ out of every 16 individuals and are grey-coloured because they possess the agouti gene irrespective of whether they also possess the black gene. The one chocolate-coloured rat is homozygous for both the recessive genes [a] and [b]. Thus an apparently new type is produced by the bringing together of recessive genes from different parent strains.

A test cross of the F_1 hybrids with a doubly recessive chocolate-coloured rat would produce the ratio 2 agouti : 1 black : 1 chocolate.

(iii) If the original agouti rat is $\dfrac{A}{A}\dfrac{B}{b}$, then the cross with the black rat $\dfrac{a}{a}\dfrac{B}{B}$ produces offspring which are all agouti in phenotype but are of two different genotypes.

Thus:

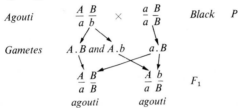

The result of inbreeding these F_1 rats depends on which type is mated with which, and as there is no difference in external appearance the results are liable to be confusing. It is thus obviously better to use the test cross, in which case the rats of genotype $\dfrac{A}{a}\dfrac{B}{B}$ have equal numbers of agouti and black offspring while those of genotype $\dfrac{A}{a}\dfrac{b}{B}$ produce 2 agouti : 1 black : 1 chocolate offspring.

Whenever there is **linkage** instead of independent assortment, the results are totally different from the above. As in the last case quoted, test-crossing gives a much clearer picture of events than simple inbreeding, because it indicates the ratio in which the different possible types of gametes are actually being produced.

In the rat the curly hair gene is linked with that for blackness.

Thus when a smooth-haired chocolate-coloured rat $\dfrac{cu}{cu}\dfrac{b}{b}$ is crossed with a curly-haired black rat $\dfrac{Cu}{Cu}\dfrac{B}{B}$, the F_1 hybrids are all curly haired and black, i.e. they show both dominant characters. But when these rats are test-crossed, there are more of the two original parental types than of the other two types, irrespective of sex. The linkage is, however, imperfect, and the **crossover types** occur in constant ratio to the **non-crossover types**.

Note. If the original parents had been a smooth-haired black rat and a curly-haired chocolate-coloured rat, then such would have been the non-crossover types, while smooth-haired chocolate-coloured and curly-haired black rats would have been produced in smaller members as crossover types.

CHROMOSOMES

The observed behaviour of the **genes** or factors governing the hereditary characteristics can be explained on the assumption that the genes are borne in linear order on the chromosomes, i.e. the darkly-staining bodies found in the nuclei of all living cells. The genes themselves may be considered as specialized regions of the **chromosomes**, distinctive banding of which has been observed in certain cases, e.g. the salivary gland chromosomes of the fruit-fly, *Drosophila*.

During ordinary cell division or **mitosis**, each chromosome divides longitudinally and the halves separate so that the nuclei of the two daughter cells have identical constitutions. During gamete formation the first maturation division is the **reduction division** or **meiosis** in which the chromosomes pair and one member of each pair passes into each daughter cell. Thus the members of each allelomorphic pair of genes are separated and pass into different gametes. This produces the exclusion effect noted in Mendel's first law. The chromosome pairs segregate independently, so that genes borne on different chromosome pairs are independently assorted as postulated in Mendel's second law. Genes borne on the same chromosome pair, however, tend to be carried through meiosis together, thus producing the effects of linkage. The imperfection of linkage is due to the observable phenomenon known as **crossing over**, which occurs during the early stages of meiosis. After pairing up, each of the chromosomes divides longitudinally, forming two **chromatids**. The chromatids twist round one another and occasionally break. Wherever a break occurs it produces a similar break in the corresponding position of another chromatid. On re-joining, the pieces are liable to become exchanged. If the exchange is between two chromatids from the same chromosome, the final genetic effect is nil, but if it is between chromatids from different chromosomes, then the genes on opposite sides of the break change partners and crossover types result.

With a few exceptions it has been found that the position of the crossover is indiscriminate therefore the frequency of crossing over between two given linked genes is an indication of the relative distance between their loci. Thus by linkage and crossing-over experiments it is possible to construct **chromosome maps**.

In the case of the rat and other mammals, chromosome mapping is difficult because insufficient characters are known and because of the difficulty in obtaining significant statistical results where families are small and generations relatively long.

The rat has twenty-one pairs of chromosomes: therefore, twenty-one linkage groups are possible, but so far only six are known.

A map has been made of part of the so-called albino chromosomes, i.e. the chromosomes that carry the alleles [C] and [c].

It is, of course, possible that the whole set of loci will have to be shifted along the map if any alleles are found to lie on the far side of the lethal [l].

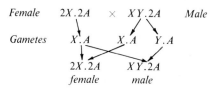

The occurrence of allelomorphic genes is due to changes or **mutations** of the original or normal genes of the wild type individuals. The full list of causes of such mutations is unknown, though the mutation rate has been increased experimentally by subjecting the germinal tissue to X-rays in a number of different species.

A mutation may be dominant, e.g. [Cu] for curly hair, or recessive, e.g. [b] for chocolate colouring instead of black, and [w] for waltzing instead of normal gait. In nature, any mutation which is advantageous to the individual in its struggle for survival is likely to be perpetuated and thus lead to the evolution of new races and eventually new species. Any dominant mutation which is disadvantageous is likely to be rapidly wiped out, while any disadvantageous recessive mutation may be carried hidden for many generations and only show its effects sporadically.

Occasionally successive mutations occur at the same locus, producing a **multiple allelomorphic series** instead of a simple allelomorphic pair. When this happens there is serial dominance. Thus in the rat full colour [C] is dominant to both ruby-eyed [c^r] and true albino [c], while ruby-eyed is dominant to true albino though recessive to full colour, and true albino is recessive to both ruby-eyed and full colour. A similar set of multiple allelomorphs is the self-coloured, Irish, hooded series, [H], [h^i] and [h].

Sex in mammals is governed genetically by the **sex chromosomes** known as X and Y. The female mammal is homozygous for sex and has two X chromosomes. The male mammal is heterozygous for sex and has one X and one Y. The non-sex chromosomes are known collectively as **autosomes**. Thus the female has [2X.2A] in each of its somatic cells and [X.A] in each of its gametes, while the male has [XY.2A] in each of its somatic cells, [X.A] in half of its gametes and [Y.A] in the other half. Chance mating gives rise to approximately equal numbers of male and female offspring, thus:

Female $2X.2A$ × $XY.2A$ Male

Gametes $X.A$ $X.A$ $Y.A$

$2X.2A$ $XY.2A$
female   male

In man, certain characters have been found to be **sex-linked**, e.g. colour blindness. The dominant and recessive alleles can occur on the X chromosomes but the Y chromosomes behave as if they carried recessive genes only and are in fact complete or almost complete dummies. No sex-linked characters have so far been discovered in the rat.

There is no evidence of **cytoplasmic inheritance** in rats or mice, though in the latter a number of maternal effects are known to influence the development of the young irrespective of their genetical constitution.

Analysis of the material of the chromosomes has shown that the essential substance of heredity in all living organisms is **deoxyribonucleic acid (DNA)**. This substance is composed of long chains of nitrogen-containing bases (two purines, **adenine** and **guanine**, and two pyrimidines, **thymine** and **cytosine**) joined together by **sugar-phosphate** links. The varied order of the bases along the chain produces differences in the DNA and thus in the chromosomes. Each chain is capable of exact self-replication, an essential feature of the behaviour of chromosomes during cell division. The DNA molecules act as templates for the production of **ribonucleic acid (RNA)** the specificity of which is responsible for the specificity of structure of the enzymes and other proteins manufactured in the cells. The genes can be considered to represent comparatively short portions of the DNA molecules and the alleles to represent variations in these portions which in turn produce variations in the cell proteins and thus variations in physiology and structure. Breaks in the chromosomes, i.e. in the DNA molecules, resulting in **deletions** (losses), **reduplications** (repeats), **inversions** (turning round of sections) and **translocations** (exchanges between chromosomes) may show themselves as mutant effects or may simply affect segregation of the chromosomes at meiosis and thus affect fertility. X-rays and atomic radiation are particularly potent in bringing about such breakages.

BIBLIOGRAPHY

Reference has been made to numerous publications, including:

BEST AND TAYLOR, *The physiological basis of medical practice*
BULLOUGH, *Vertebrate sexual cycles*
CLEGG AND CLEGG, *Biology of the mammal*
CROFT, *An introduction to anaesthesia of laboratory animals* (U.F.A.W.)
FARRIS, *The care and breeding of laboratory animals*
FARRIS, *The rat in laboratory investigation*
GREENE, *Anatomy of the rat* (Trans. Amer. Phil. Soc., vol. 27)
GRUNEBERG, *The linkage relations of a new lethal gene in the rat* (*Rattus norvegicus*) (Genetics, vol. 24, 1939)
HAGGIS, MICHIE, MUIR, ROBERTS AND WALKER, *Introduction to molecular biology*

KARLSON, *Introduction to modern biochemistry*
KING AND CASTLE, *Linkage studies of the rat* (*Rattus norvegicus*) (Proc. Nat. Acad. Sc., vol. 21, 1935)
RUCH AND PATTON, *Physiology and biophysics*
SCOTT, *Animal behaviour*
THORPE, *Learning and instinct in animals*
VINTER, *Kind killing*
VOLRATH, *Animals in schools*
WITSCHI, *The development of vertebrates*
WORDEN AND LANE-PETTER, *U.F.A.W. Handbook on the care and management of laboratory animals*
Chambers Encyclopaedia
Encyclopaedia Britannica

INDEX

Note: Names of individual bones are omitted